LOOPY LOGIC PROBLEMS

& OTHER PUZZLES

IVAN MOSCOVICH

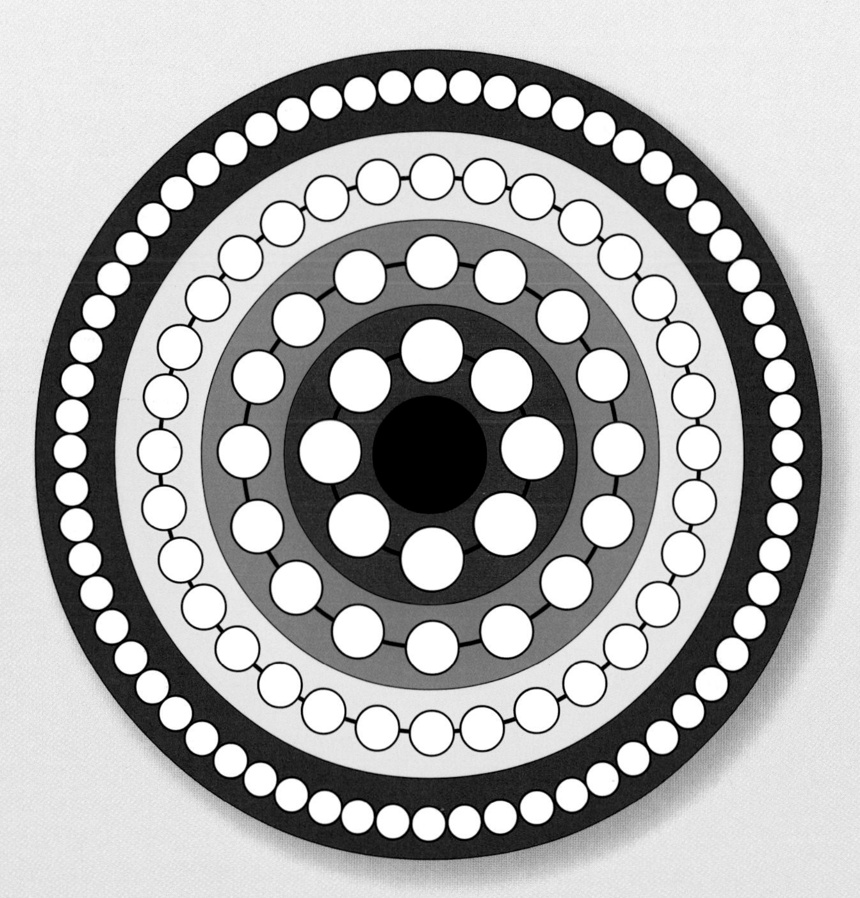

DOVER PUBLICATIONS, INC.
MINEOLA, NEW YORK

To Anitta, Hila, and Emilia, with love

Copyright

Bibliographical Note

This Dover edition, first published in 2012, is an unabridged republication of the work originally published in 2006 by Sterling Publishing Company, Inc., New York

International Standard Book Number
ISBN-13: 978-0-486-49069-4
ISBN-10: 0-486-49069-6

Manufactured in the United States by Courier Corporation
49069601
www.doverpublications.com

Contents

Also available from Dover in
Ivan Moscovich's Mastermind Collection:

Leonardo's Mirror & Other Puzzles
The Monty Hall Problem & Other Puzzles

Introduction

Ever since my high school days I have loved puzzles and mathematical recreational problems. This love developed into a hobby when, by chance, some time in 1957, I encountered the first issue of *Scientific American* with Martin Gardner's mathematical games column. And for the past 50 years or so I have been designing and inventing teaching aids, puzzles, games, toys, and hands-on science museum exhibits.

Recreational mathematics is mathematics with the emphasis on fun, but, of course, this definition is far too general. The popular fun and pedagogic aspects of recreational mathematics overlap considerably, and there is no clear boundary between recreational and "serious" mathematics. You don't have to be a mathematician to enjoy mathematics. It is just another language, the language of creative thinking and problem-solving, which will enrich your life, like it did and still does mine.

Many people seem convinced that it is possible to get along quite nicely without any mathematical knowledge. This is not so: Mathematics is the basis of all knowledge and the bearer of all high culture. It is never too late to start enjoying and learning the basics of math, which will furnish our all-too sluggish brains with solid mental exercise and provide us with a variety of pleasures to which we may be entirely unaccustomed.

In collecting and creating puzzles, I favor those that are more than just fun, preferring instead puzzles that offer opportunities for intellectual satisfaction and learning experiences, as well as provoking curiosity and creative thinking. To stress these criteria, I call my puzzles Thinkthings.

The *Mastermind Collection* series systematically covers a wide range of mathematical ideas, through a great variety of puzzles, games, problems, and much more, from the best classical puzzles taken from the history of mathematics to many entirely original ideas.

A great effort has been made to make all the puzzles understandable to everybody, though finding some of the solutions may be hard work. For this reason, the ideas are presented in a highly esthetic visual form, making it easier to perceive the underlying mathematics.

More than ever before, I hope that these books will convey my enthusiasm for and fascination with mathematics and share these with the reader. They combine fun and entertainment with intellectual challenges, through which a great number of ideas, basic concepts common to art, science, and everyday life, can be enjoyed and understood.

Some of the games included are designed so that they can easily be made and played . The structure of many is such that they will excite the mind, suggest new ideas and insights, and pave the way for new modes of thought and creative expression.

Despite the diversity of topics, there is an underlying continuity in the topics included. Each individual Thinkthing can stand alone (even if it is, in fact, related to many others), so you can dip in at will without the frustration of cross-referencing.

I hope you will enjoy the *Mastermind Collection* series and Thinkthings as much as I have enjoyed creating them for you.

—Ivan Moscovich

Gears make easy work of heavy loads and almost every machine uses these simple devices in some way. Try to determine how the various weights in this puzzle will be affected by turning the bottom red gear.

▲ UP OR DOWN?

If you turn the bottom red gear counterclockwise, what will happen with the four numbered weights?

Which will go up and which will go down?

ANSWER: PAGE 98

▲ SNEEZE

When sneezing, people close their eyes for about half a second. Imagine you sneeze while driving your car at 65 miles per hour. A driver in a car about 60 feet ahead of you suddenly brakes to avoid a collision with a cat crossing the road in front of him.

How far will you have traveled with your eyes closed before you too can brake and stop your car?

Can an accident be avoided?

Hint: there are 1760 yards in a mile.

ANSWER: PAGE 98

Speed, velocity, and acceleration

Understanding speed, velocity, and acceleration is fundamental to modern life. Speed is the distance covered in a unit of time. The units could be mph or ft/s (miles per hour, or feet per second). Speed measurement is always relative.

Velocity is the speed of an object in a particular direction. If the speeds of two cars are given, it will tell you nothing about whether they will collide. Giving their directions as well will tell you whether they will.

Acceleration is the rate of change of velocity. It tells how quickly an object changes its speed (or velocity) every second. The acceleration due to the earth's gravity pull is 30 ft/s/s, which means that an object dropped increases its speed by 30 ft/s every second it falls until it reaches terminal velocity. In a plane or subway you can only tell that you are moving if your speed changes—in other words, when you accelerate.

1 5 10 15 20 25 30 35 40 45 50 55 60 65 70 75 80 85 90

▲ INCLINED PLANE

Let's say we release a ball on an inclined plane and mark its position after exactly 1 second of descent. We then divide the whole length of the slope into units of this length as shown. Can you mark the positions of the ball after 2, 3, 4, 5, 6, 7,8, and 9 seconds?

Will these marks change if the slope is steeper?

Galileo used the inclined plane extensively for his famous experiments with falling bodies, since the motion of a body on an inclined plane is similar to that of a body in free fall, except that its velocity is slowed by the slope, for easier observation and measurement.

ANSWER: PAGE 98

▼ RACE

Each runner competing in the 100-yard race ran the entire race at a uniform speed.

Runner A crossed the finish line 10 yards ahead of runner B, who crossed the finish line 10 yards ahead of runner C.

By how much did runner A beat runner C?

ANSWER: PAGE 98

Motor-racing drivers and pilots must be able to react to situations in split seconds. But good reflexes are also needed in many everyday life situations, such as playing sports. Try this reaction test on your friends.

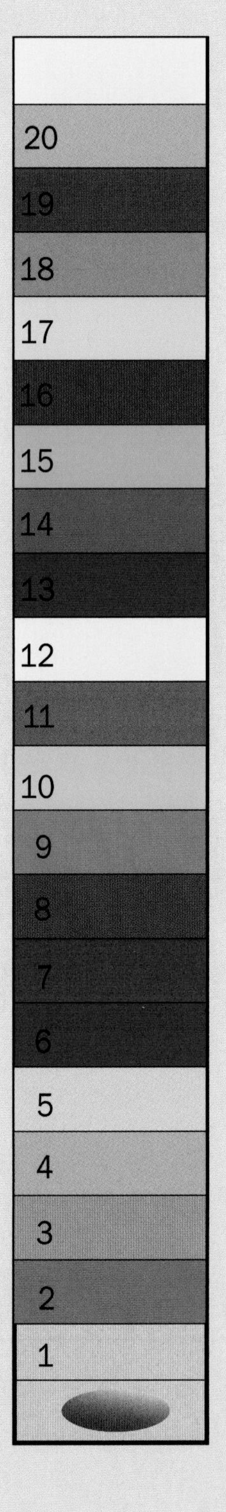

▲ REFLEX RULER REACTION

Hold a ruler with one hand on top and the other hand at the bottom with open fingers not quite touching the ruler.

When you release the ruler it won't be difficult to close your fingers and catch it. But try the same experiment with your friends and see whether they can catch the ruler easily when you release it.

It will not be easy for them to catch the ruler.

Why?

ANSWER: PAGE 98

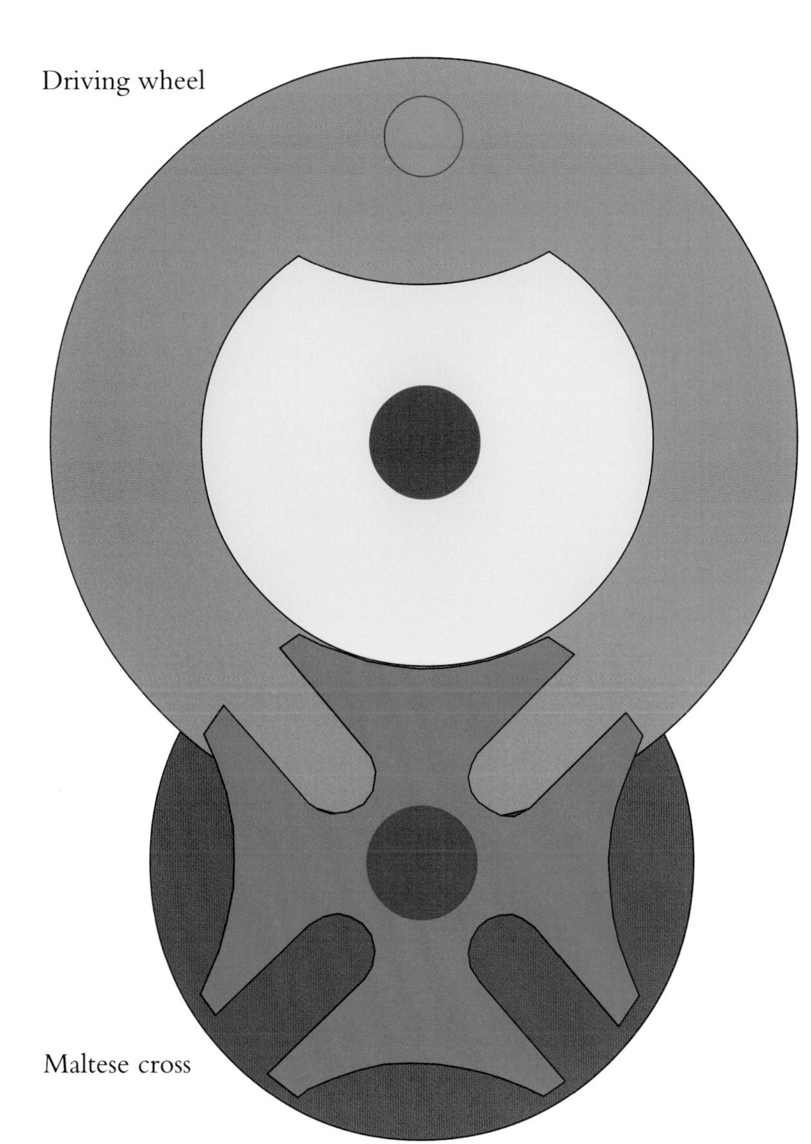

Basic mechanisms

Schematic models of machines and basic mechanisms can help us visualize the basic principles of their operation.

Ordinarily when we observe geometric constructions with simple shapes, such as triangles, squares, and circles, we think of them as being static, each with a definite dimension that does not change. Geometry, however, also makes use of the concept of motion. The field of geometry concerned with motion is known as kinematics.

▶ MALTESE CROSS MECHANISM

The Maltese cross mechanism is the chief mechanism of many tools and mechanisms.

What is the machine on the right? How does it work and why does the Maltese cross help?

ANSWER: PAGE 98

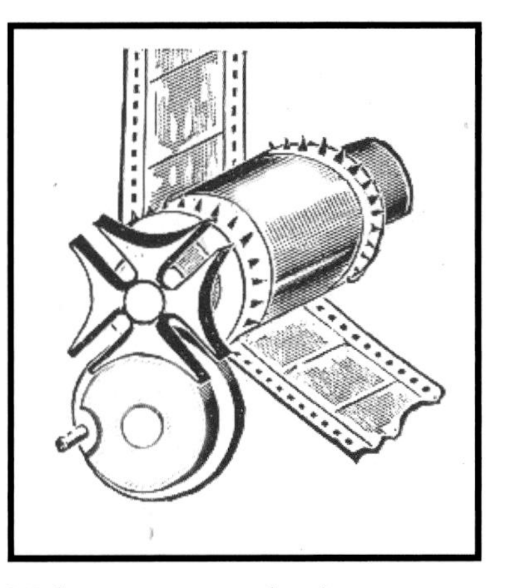

Maltese cross mechanism

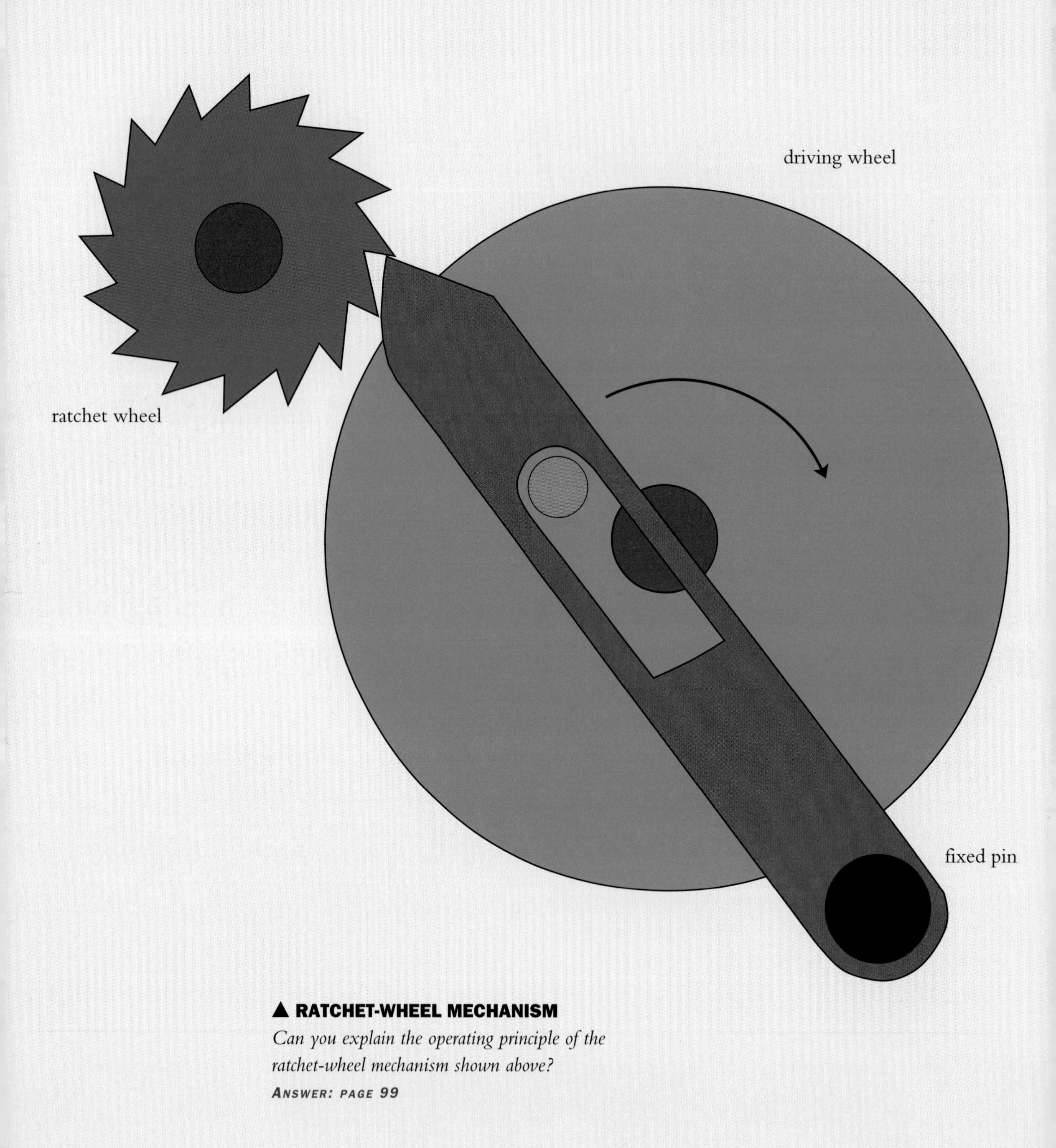

▲ RATCHET-WHEEL MECHANISM

Can you explain the operating principle of the ratchet-wheel mechanism shown above?

ANSWER: PAGE 99

▲ SLIDING-FRAME MECHANISM

Can you explain the operating principle of the sliding-frame mechanism shown above?

Answer: page 99

The swing of a pendulum is a fascinating phenomenon to watch and has intrigued scientists from Galileo to Einstein. The next few puzzles ask questions about how and why pendulums work the way they do.

PENDULUMS

Pendulums have been the object of study for centuries. A pendulum can keep time, measure the force of gravity, and sense relative motion. Galileo realized that a pendulum swings for almost the same length of time with either a small or a large swing. With this simple observation he invented the pendulum clock. The mass of the bob (the weight on the end of the string) does not matter, but as the length is increased the period lengthens by the square root of the length of the pendulum.

What happens to pendulums on top of mountains or on the Moon, where gravity is weaker? In such places, pendulums are slower. And a pendulum on a massive planet such as Jupiter would swing faster.

But if a pendulum swings faster under stronger gravitational pull, why doesn't a heavier bob swing faster than a light bob? The reason is that the increased inertia of the heavier bob requires more force to accelerate it, and the extra force required is exactly the same as the increased gravitational attraction. This means that inertia and gravity are precisely related. For Newton, this exact relationship was a total mystery. He believed that it was too much of a coincidence, and yet there was no visible or conceptual link between inertia and gravity to explain it.

This was one of the questions that led Einstein to theorize that inertia and gravity are ultimately the same thing. But to say this, he had to create a new picture of the universe—and then persuade his fellow physicists to follow him, which they ultimately did.

So, much hangs on a pendulum.

Galileo's pendulum

The principle of the pendulum clock arose from Galileo's work on pendulums. Galileo showed that the period of a pendulum is independent of its weight, a property known as "isochronism." This property permits its use in clocks for regulating the timing mechanism.

Galileo described a pendulum clock in 1641, one of the earliest designs to control the rotary motion of a spindle by the swinging movement of a pendulum.

The foundations of the modern clock-maker's art were laid down by the Dutch scientist Christian Huygens in *The Clock*, published in 1658. Huygens constructed an accurate timekeeping instrument by applying a pendulum to a clock driven by weights. Weights kept the clock going, but the pendulum regulated the rate of the movement.

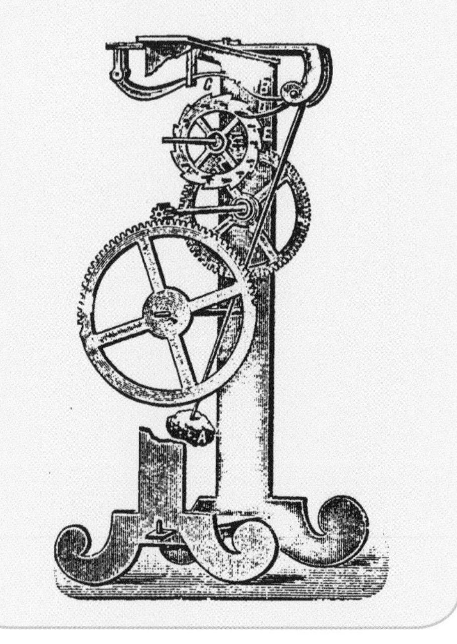

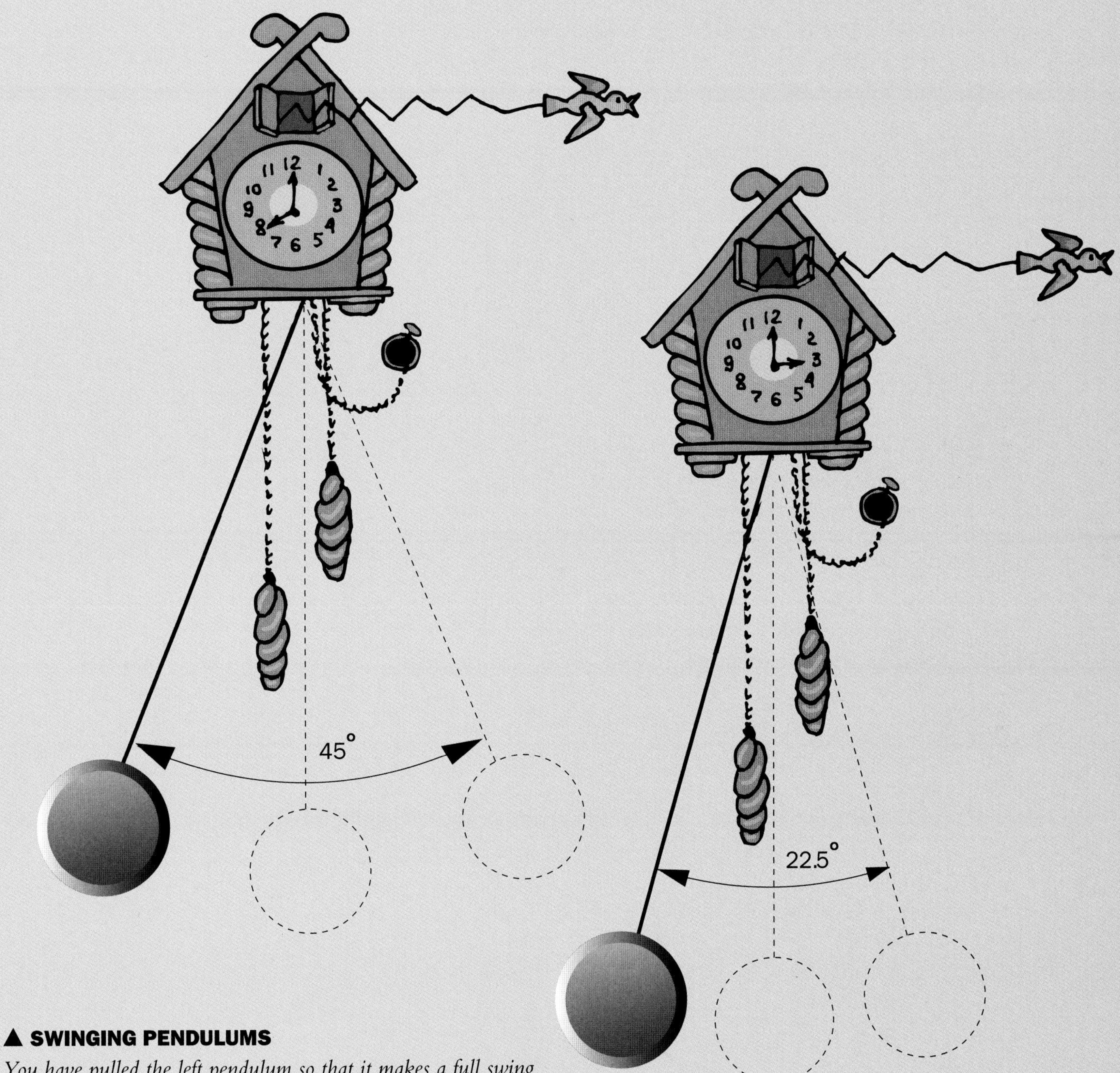

▲ SWINGING PENDULUMS

You have pulled the left pendulum so that it makes a full swing of about 45 degrees. You have timed it and it takes 5 seconds for a full swing. In a second experiment you pulled the pendulum so that its swing is only 22.5 degrees as shown. How long will a full swing take this time?

By the way, can you tell what is wrong in this problem?

ANSWER: PAGE 99

◀ SIMPLE HARMONIC MOTION

By attaching a pen to the end of a swinging pendulum and registering the path traced by the pen on a rolling paper as shown, we will get a characteristic curve.

Can you visualize the shape of the resulting curve before it appears on the paper?

ANSWER: PAGE 100

Pendulum

One of the examples of simple harmonic motion is the pendulum. A pendulum swings to and fro, alternately accelerated and then retarded by earth's gravity. If there were no friction or air resistance, the pendulum would lose no energy and would swing endlessly.

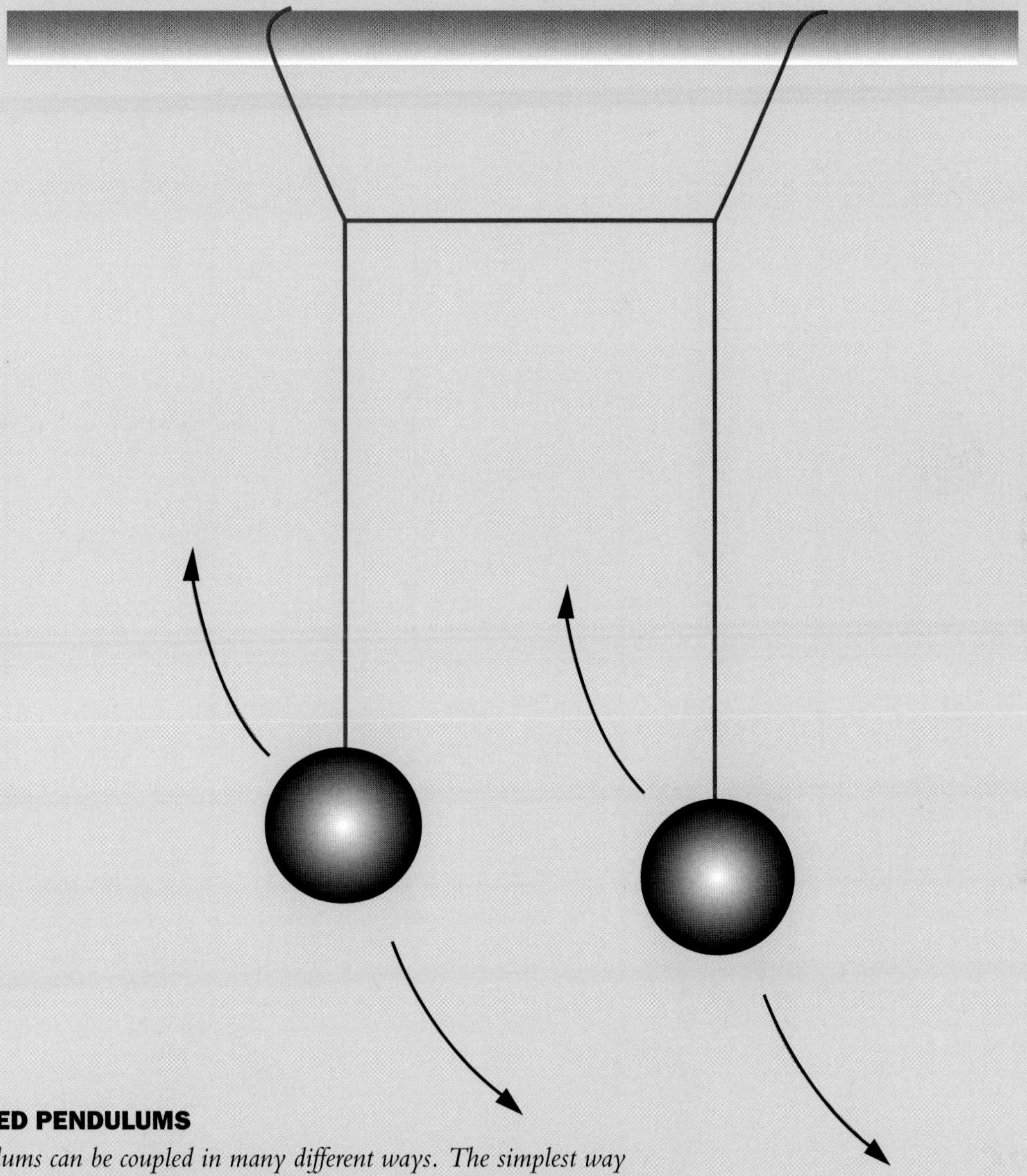

▲ COUPLED PENDULUMS

Two pendulums can be coupled in many different ways. The simplest way is with strings hung from a dowel, as shown here. You can make your own version by using two beads for the pendulums and hanging them from a long pencil. The pendulums are tied to a string that connects them. The pendulums can then swing perpendicular to the connecting string.

Can you predict what will happen to the system as a whole if you swing one of the pendulums only?

ANSWER: PAGE 100

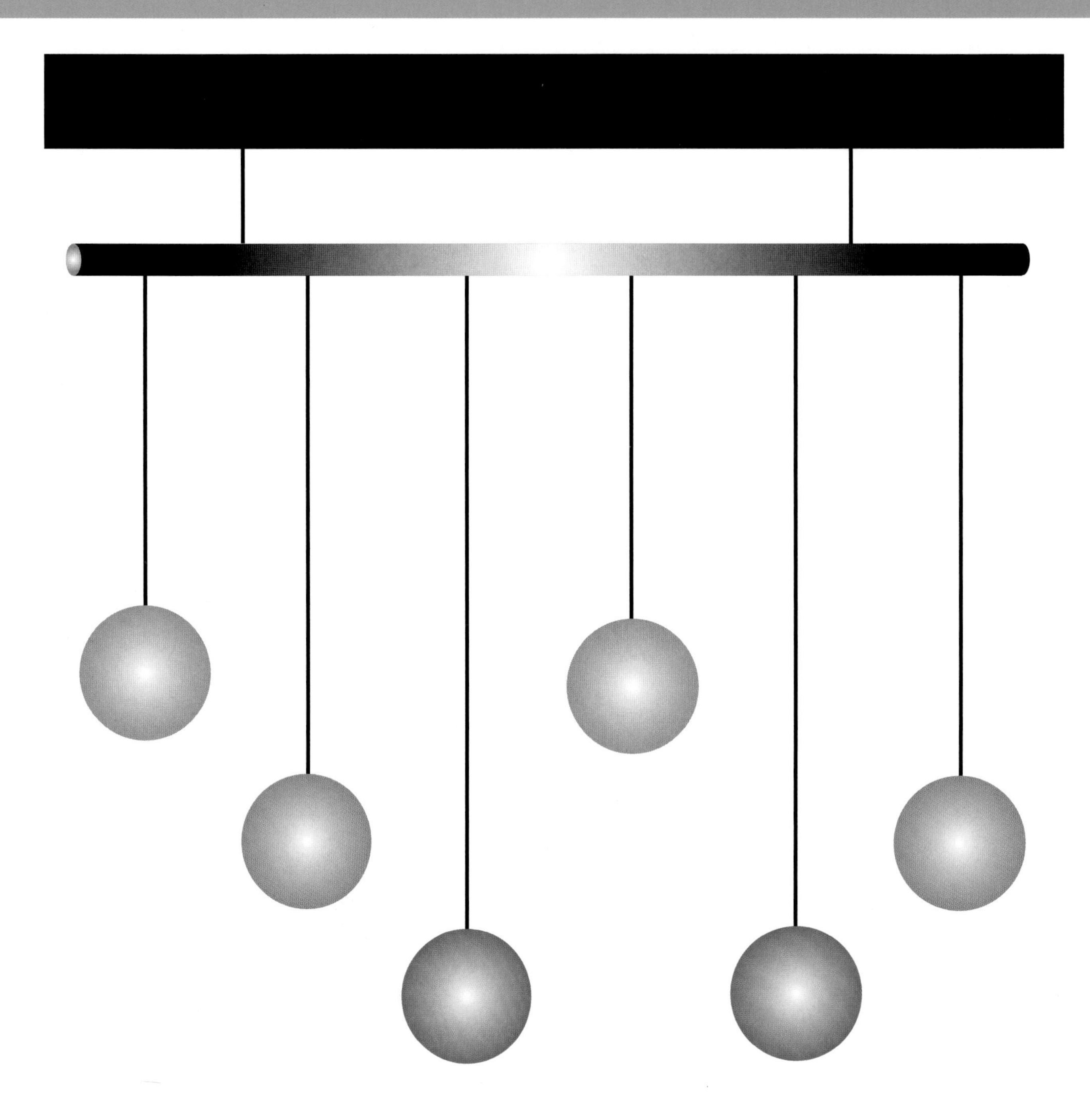

▲ RESONANCE

The three pairs of pendulums are coupled by a horizontal bar as shown. The identically colored pendulums have identical lengths.

Start swinging any one of the six pendulums. The bar transmits impulses to all the other pendulums. What will happen after a while?

ANSWER: PAGE 101

▼ ONE-TON PENDULUM

Can you explain how the boy can cause the massive stationary pendulum weighing one ton to start swinging, using only a thin piece of string with a small magnet at its end?

ANSWER: PAGE 101

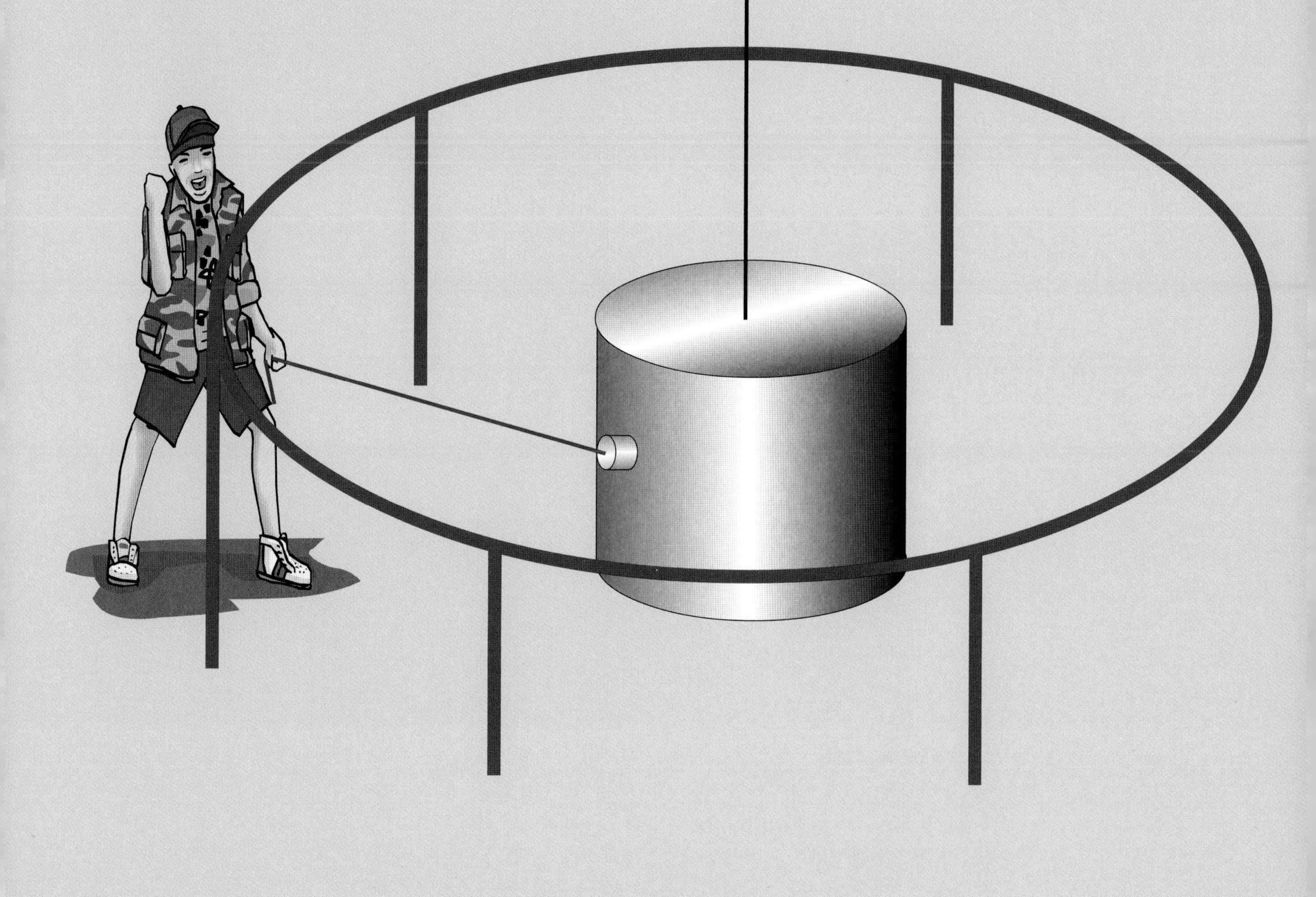

▲ FOUCAULT'S PENDULUM

How do you know that the earth moves?

Astronomers from Plato's time up until the sixteenth century tended to think that the earth sat still while everything else rotated around it. Theories contradicting this view were abundant but the problem was that convincing evidence was severely lacking.

We certainly can't feel that we are on a moving platform, but can we see the earth moving? Is it possible to watch the earth rotate?

In the year 1543, Copernicus sent a copy of his book On the Revolutions of the Celestial Orbs *to Pope Paul III with a note containing the famous understatement: "I can easily conceive that as soon as people learn that in this book I ascribe certain motions to the earth, they will cry out at once that I and my theory should be rejected".*

Some still disbelieved the theory when the French physicist Jean-Bernard Foucault was invited to arrange a scientific exhibit as part of the Paris Exhibition in 1851. From the dome of the Pantheon, Foucault hung a pendulum consisting of 66 yards of piano wire and a 59-pound cannonball. On the floor below the ball, he sprinkled a layer of fine sand. A stylus fixed to the bottom of the ball traced the path in the sand, thus recording the movement of the pendulum.

At the end an hour, the line in the sand had moved 11 degrees and 18 minutes. If the pendulum was initially swung in a straight line, how could it trace different paths in the sand?

ANSWER: PAGE 102

Rott's chaotic pendulum

You may have seen this object in science exhibits. Rott's chaotic pendulum has three pendulums connected to a freely revolving fixed structure in the form of a capital T as shown.

If a strong revolving movement is given to the T-structure, can you predict how the whole system will behave?

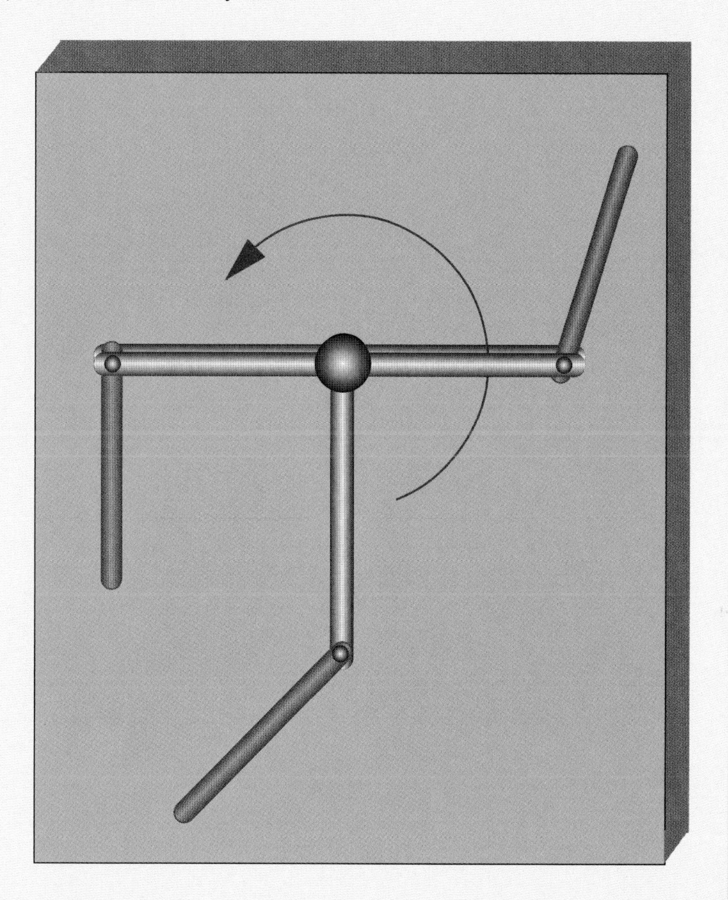

Possibly the only thing you can predict is that eventually the pendulums will stop moving.

If you swing Rott's pendulum vigorously enough that it goes full circle, it will move in unpredictable and surprising ways. The pendulum behaves in a complex, chaotic manner, depending on the precise starting conditions. Prediction is difficult, if not impossible, because there are so many variables.

Similarly, many things in nature are just as chaotic—weather, traffic flow, etc. Chaotic behaviour was first discovered by scientists investigating weather systems. They found that a tiny event—like a butterfly flapping its wings in Brazil—could determine the weather in London. So the way a small event can quickly and unpredictably get a lot bigger is known as the "butterfly effect."

Use your general mathematical knowledge to answer these two puzzles. Will your weight and the length of the staircase be a surprise or not?

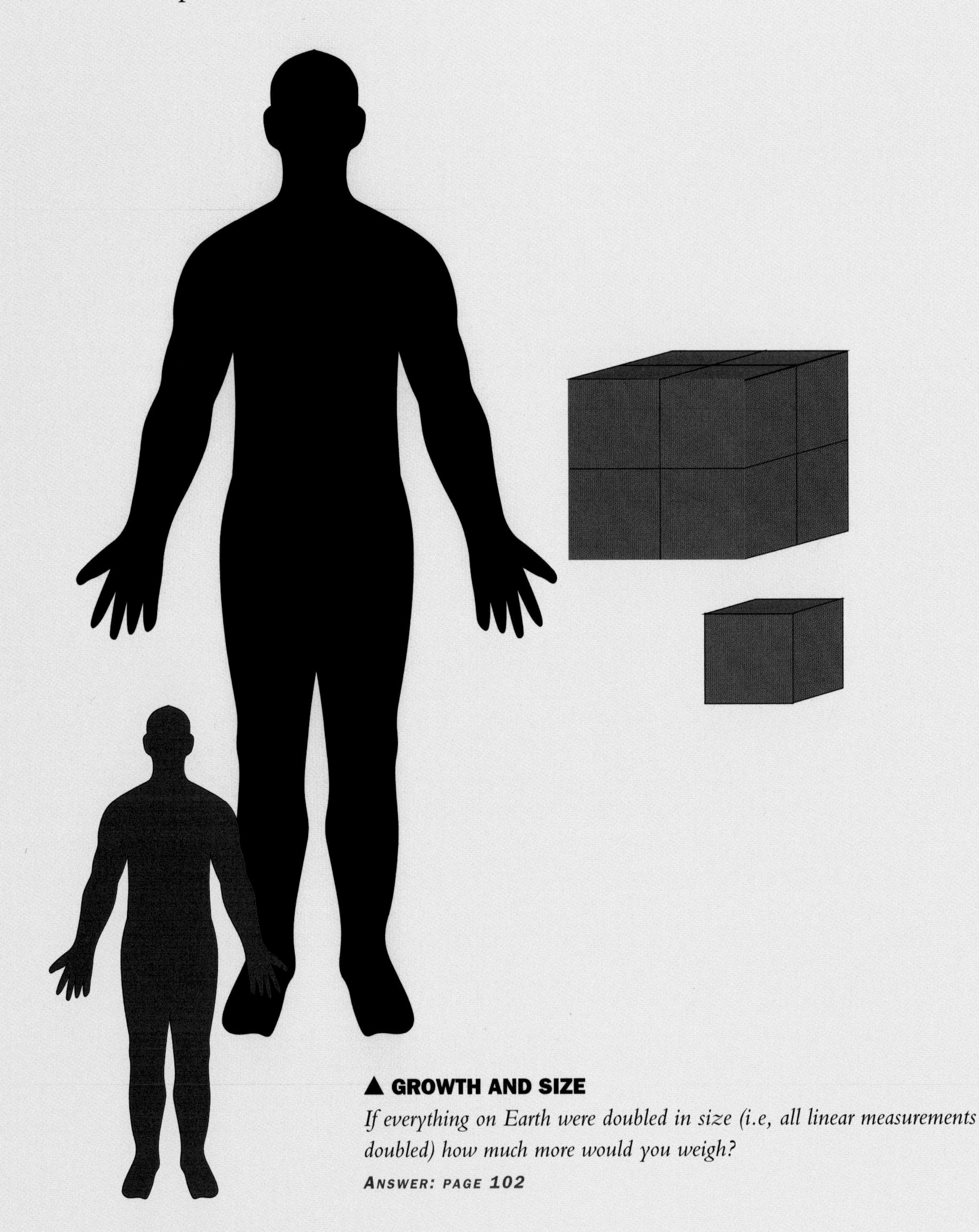

▲ GROWTH AND SIZE

If everything on Earth were doubled in size (i.e, all linear measurements doubled) how much more would you weigh?

ANSWER: PAGE 102

▼ STAIRCASE PARADOX

In the progression sequence below, how long (in units) will the length of the staircase ultimately become, if we divide the squares infinitely?

How many stairs will there be in the 10th generation?

ANSWER: PAGE 102

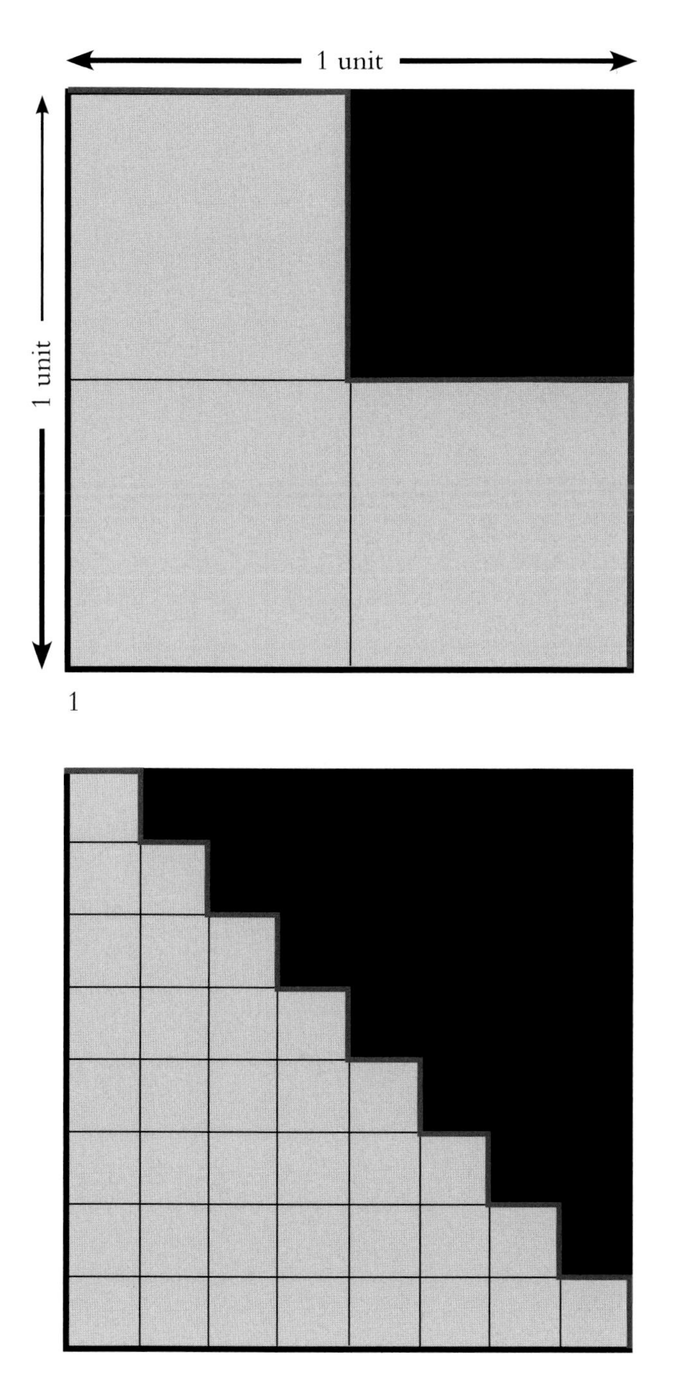

1

3

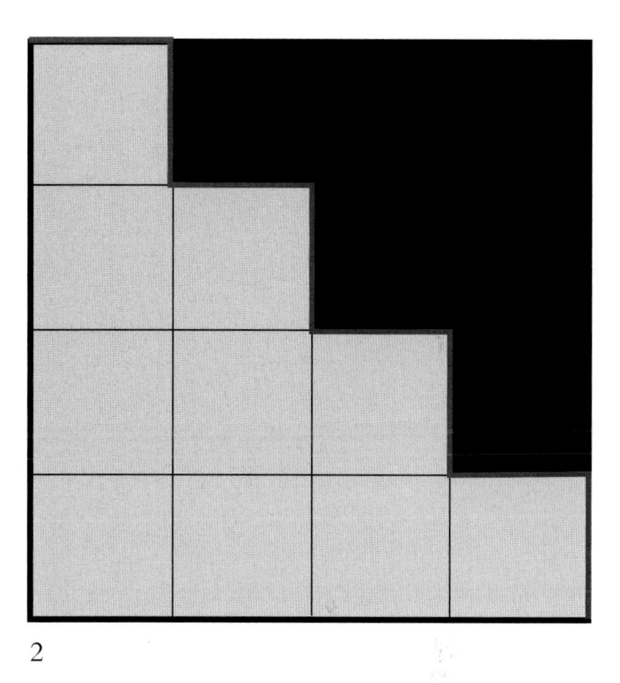

2

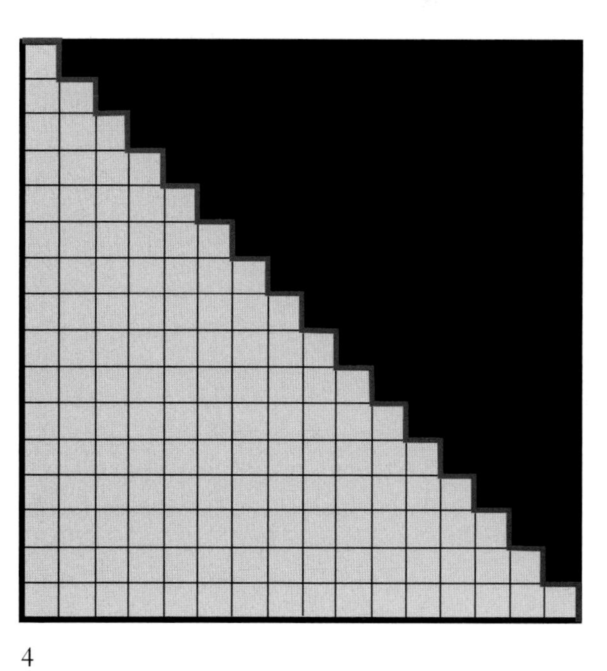

4

The concept of infinity has baffled humans for as long as we have known about the idea. What is it and how can we understand it? Try these puzzles and see if they provide a few clues.

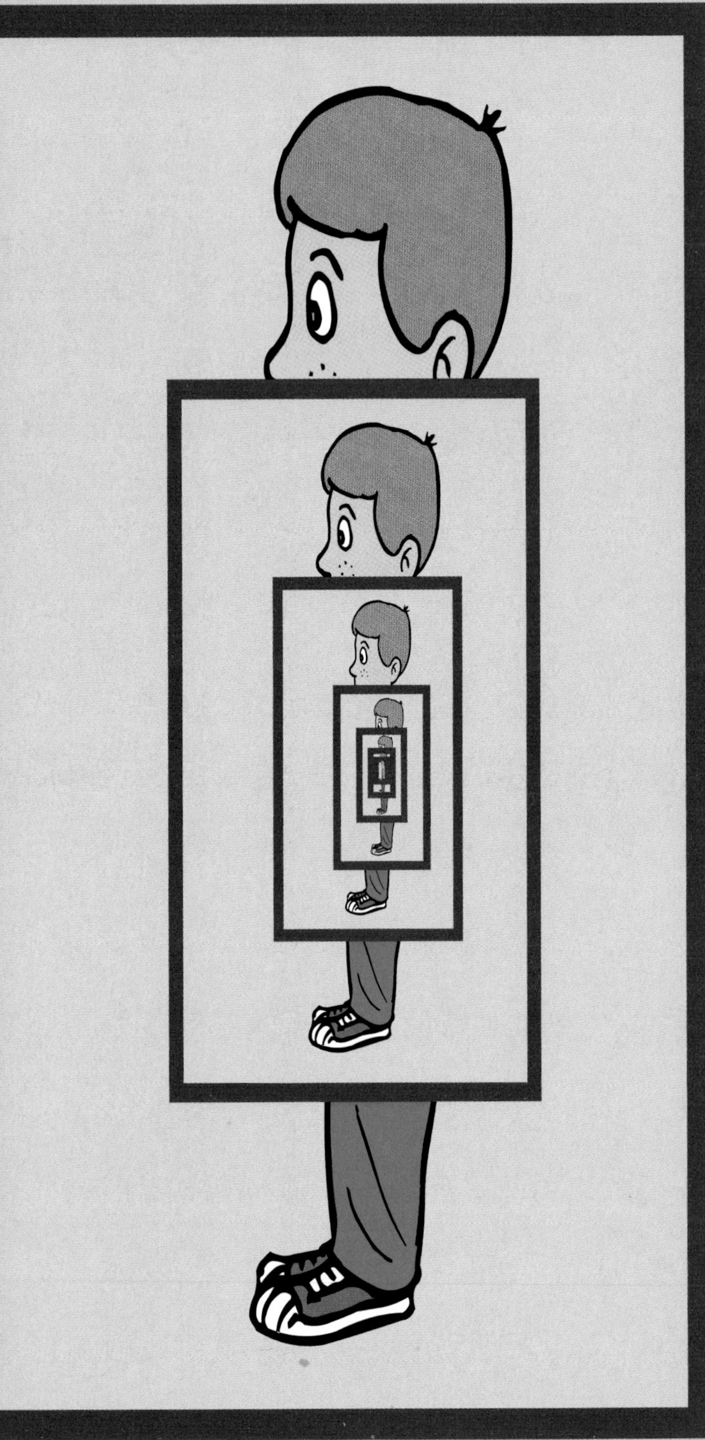

▶ INFINITY AND LIMITS

Each picture on the right is half as high and wide as the one outside it. There will be an infinite number of pictures, as you can imagine, but can you tell how high all the pictures would reach if they hung on a wall, one on top of another?

There are infinitely many boys in the pictures, but if they stood on each other's heads, the tower they made would not be infinitely high.

How high would the tower be?

ANSWER: PAGE 102

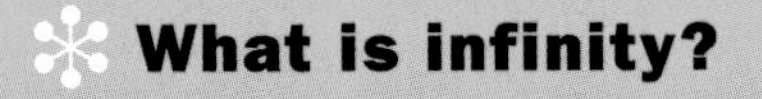

What is infinity?

Infinity is a quantity larger than anything that is fixed. It has special significance in areas such as number theory, algebra, geometry, etc.

For example, the sequence of whole numbers 1, 2, 3, 4, 5, 6, 7, and so on, is infinite, because it is always possible to add 1 to any whole number. The list is never-ending—it is infinite.

Since so many facts related to infinty are counter-intuitive, it is essential to learn its ways in order to obtain a firm grasp of the concept of limits and infinity in general.

Puzzles and paradoxes are helpful in achieving such an understanding.

c

a

b

Gnomons and the Pythagorean theorem

Hero of Alexandria defined gnomons as follows: "A gnomon is any figure which, when added to an original figure, leaves a resultant figure similar to the original."

This is nature's most common form of growth, in which the old form is contained within the new.

This is the way the more permanent tissues, such as bones, teeth, horns, and shells develop, in contrast with soft tissue, which is discarded and replaced.

(Also, the word gnomon is the term used to describe the arm on a sundial.)

◀ GNOMONIC EXPANSION

Can you figure out the relationship of the red gnomon to the Pythagorean formula:

$a^2 + b^2 = c^2$

ANSWER: PAGE 102

The puzzles on this and the following pages deal with arithmetical progressions. Test your ability to figure out questions that use numbers, skyscrapers, waterlilies, and snowflakes as subjects.

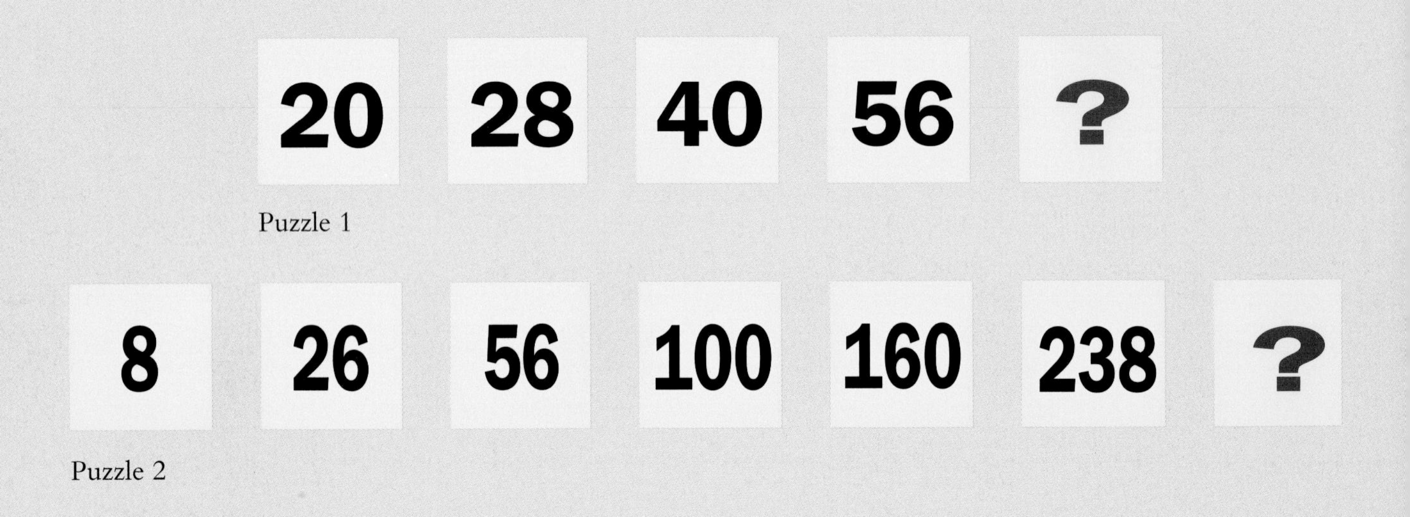

Puzzle 1

Puzzle 2

▲ ARITHMETICAL PROGRESSIONS

A sequence of numbers that has a common difference between successive pairs of numbers is called an arithmetical progression.

For example:

2 4 6 8 level 0

2 2 2 level 1

is an arithmetical progression, in which a simple multilevel analysis easily reveals the common difference of 2.

But it is not always so easy to find the common difference in arithmetical progressions, and multilevel analysis can be required.

Can you determine the next numbers in the two number sequences above?

ANSWER: PAGE 103

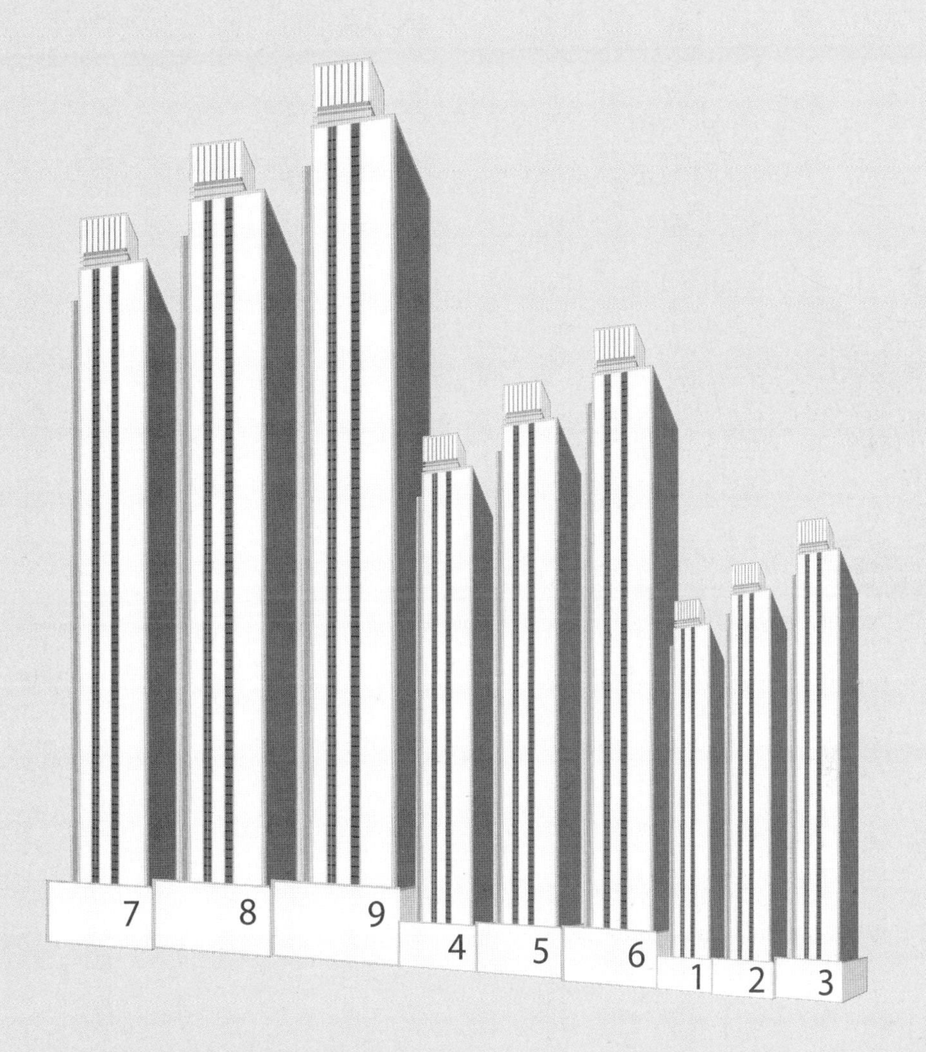

▲ ROW OF SKYSCRAPERS

The above design for a row of nine skyscrapers was rejected as boring.

For aesthetic and other reasons the clients made this request:

The buildings must be arranged in a row and each must be of a different height, but no more than three skyscrapers are allowed to go in an ascending or descending order along the row from left to right (whether they are adjacent to each other or not).

Can you find at least two arrangements of the skyscrapers to comply with these requirements?

ANSWER: PAGE 103

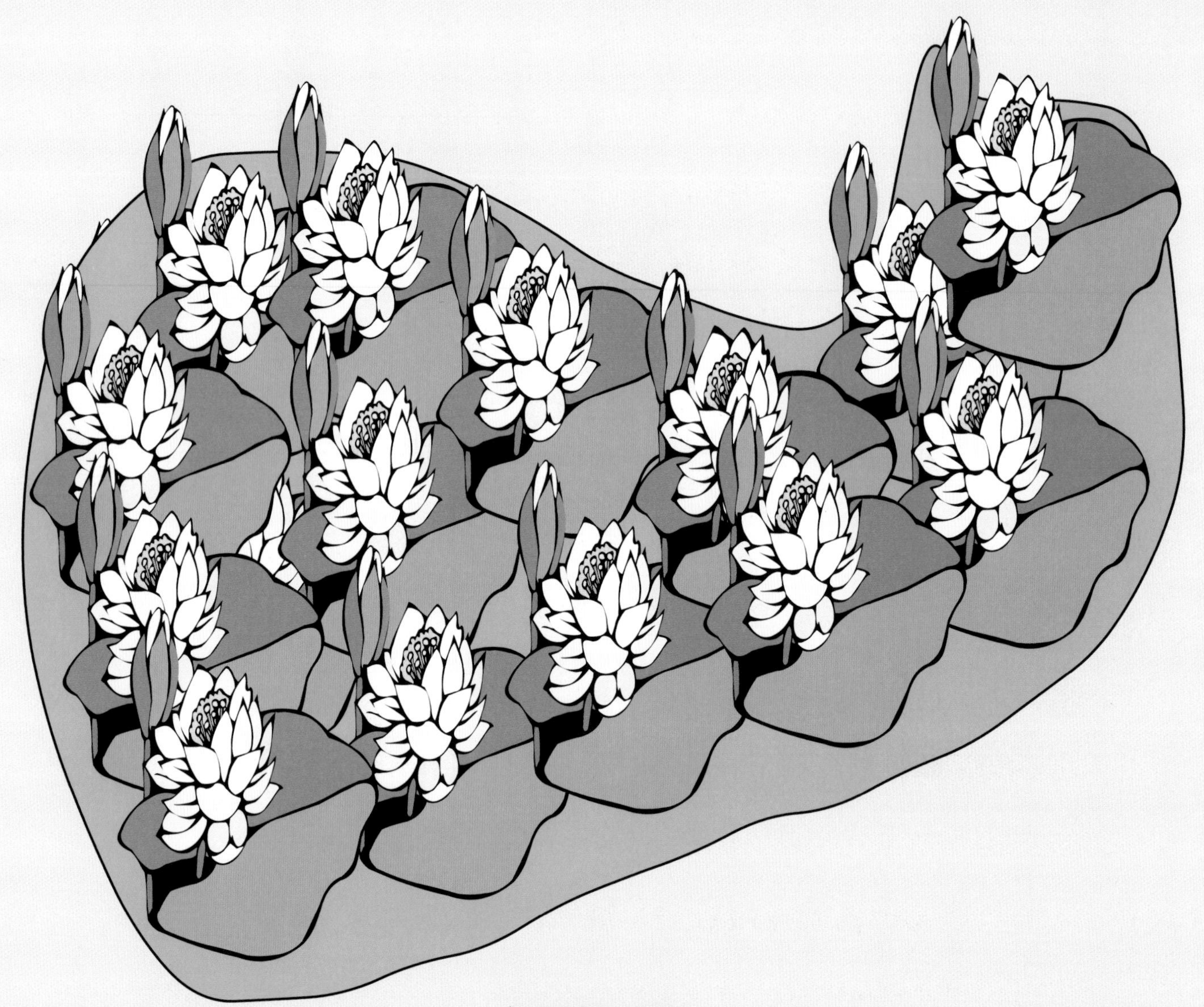

▲ WATER LILIES

The number of water lilies on the small lake doubles each day. When there is one water lily on the lake, it takes 60 days for the entire lake to be covered with water lilies.

How long would it take for two water lilies to cover the lake?

ANSWER: PAGE 104

▲ SNOWFLAKE AND ANTI-SNOWFLAKE CURVES

The four red figures above show the first four generations of the famous snowflake fractal.

The fifth generation is shown above them. The yellow area denotes the anti-snowflake curve whereby small triangles are cut out of the large triangle instead of being attached to the sides.

To make this fractal, start with an equilateral triangle whose sides have length 1. On the middle third of each of the three sides add an equilateral triangle with sides of length ⅓, and repeat the process.

Can you tell what the limit of the length of the snowflake's perimeter will be and the area it encloses as the process continues indefinitely?

Are there three-dimensional analogs of the snowflake curve?

ANSWER: PAGE 104

Here is another puzzle in which a progression rule is created. Can you work out what would happen if the process went on indefinitely?

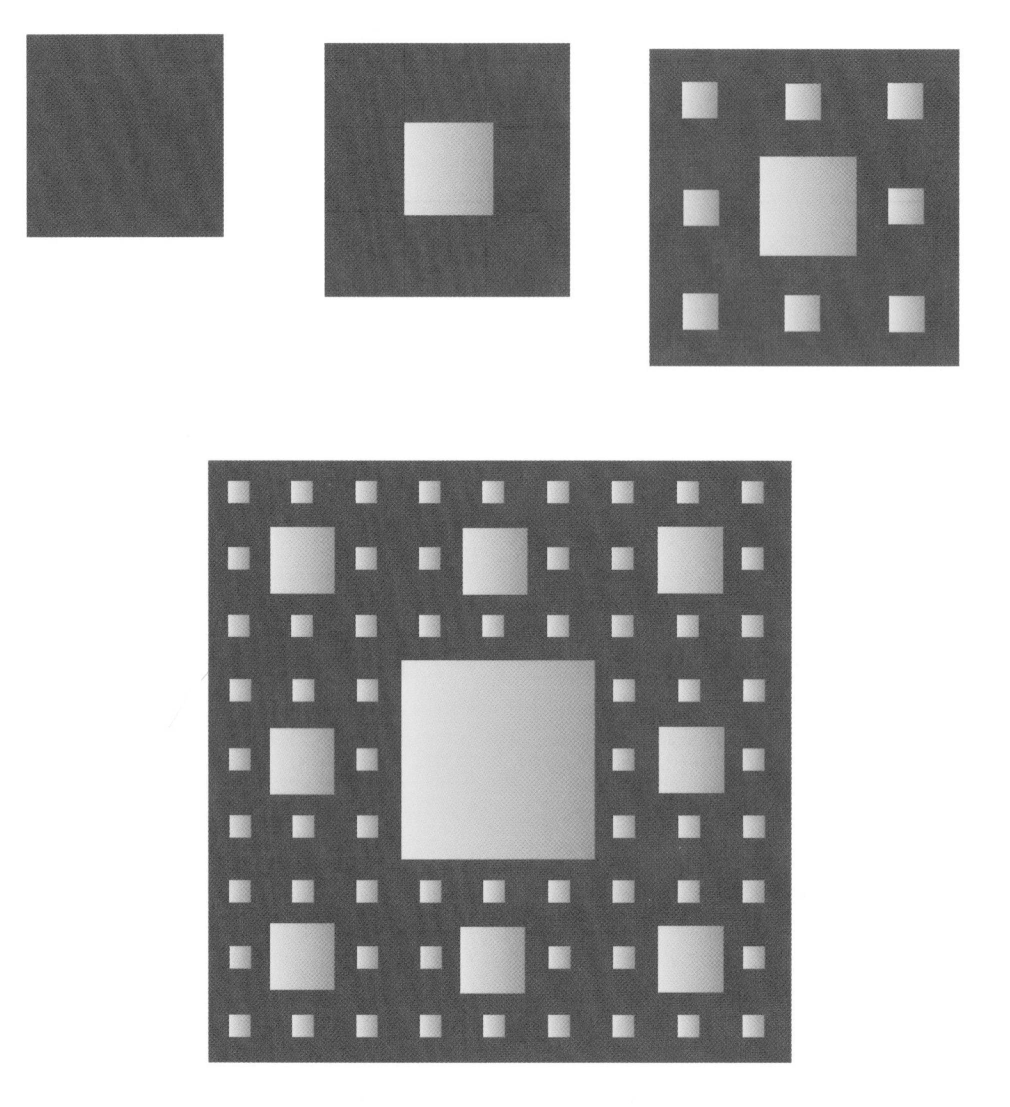

▲ SQUARES IN SQUARES

A square is divided into nine squares by dividing its side in thirds and painting the middle square gold. In the next generation, the remaining blue squares are each similarly divided, with their middle squares painted gold, and so on.

If this process goes on indefinitely, can you guess what the proportion of the gold areas in relation to the area of the initial blue square will be?

ANSWER: PAGE 104

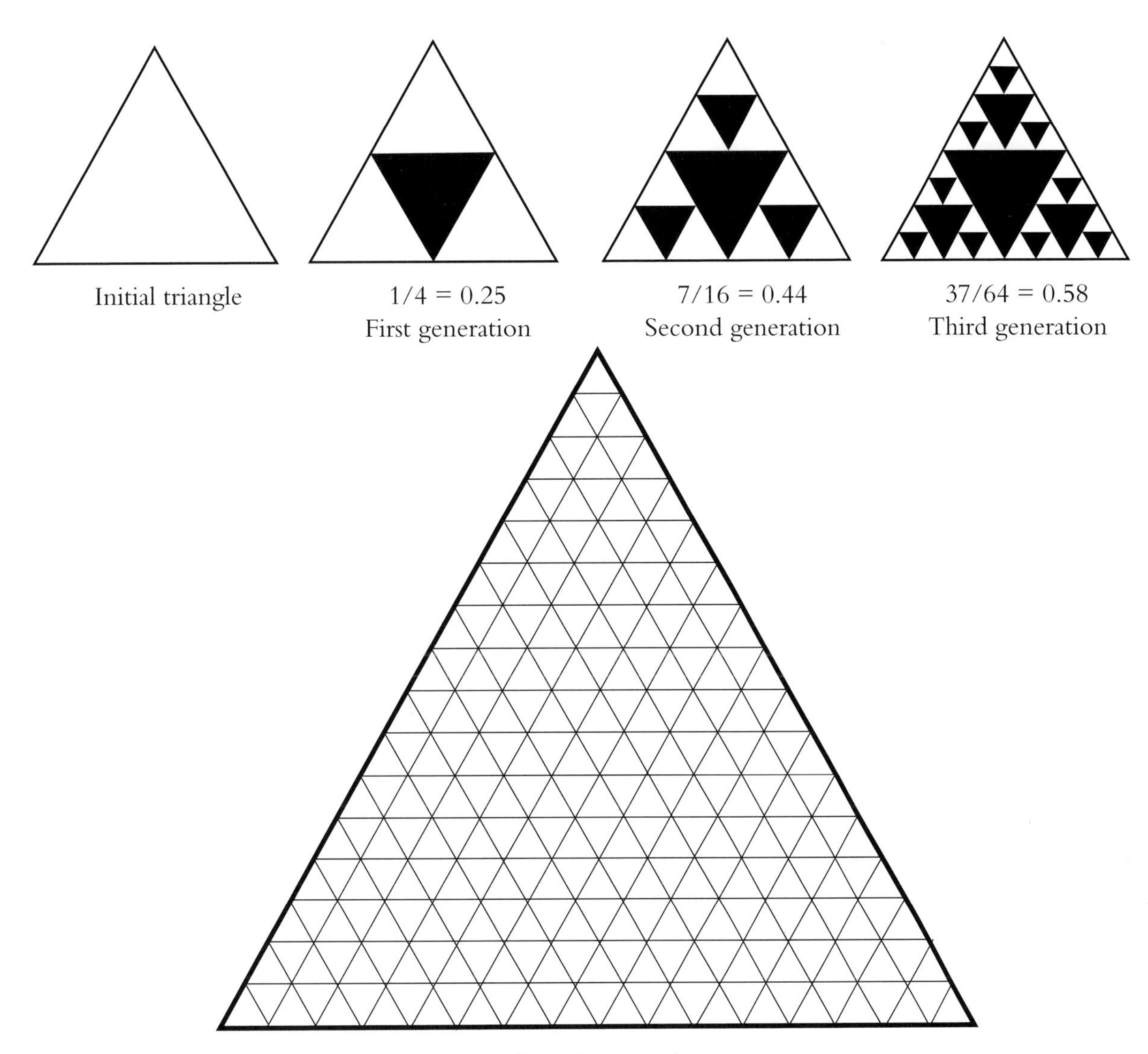

▲ SIERPINSKI TRIANGLE

Three generations of the Sierpinski triangle are demonstrated. Can you work out the fourth generation in the given triangular grid?

Can you also find the number series showing the proportions of the black areas to the area of the whole triangle in each generation?

The triangle is obtained by starting with a filled equilateral triangle, which is then divided into four smaller equilateral triangles, of which the middle one is removed, forming a triangular hole.

The three remaining filled triangles are then divided in the same way, a process which can go on indefinitely. The pattern achieved after the process reaches its limit is called the Sierpinski fractal. Vaclaw Sierpinski (1882–1969) introduced this fractial in 1916.

ANSWER: PAGE 105

Named after Italian mathematician Leonardo Fibonacci (1170–1250), the puzzles below rely on the Fibonacci series, an infinite sequence of numbers (0, 1, 1, 2, 3, 5, 8, and so on) in which each number is the sum of the previous two numbers.

First generation
1, 1...

Second generation
1, 1, 2 ...

Third generation
1, 1, 2, 3 ...

Fourth generation
1, 1, 2, 3, 5 ...

▶ FIBONACCI-SQUARE PROGRESSIONS

Four generations of Fibonacci-square progressions are shown on this page.

Using the same rules, can you create the fifth generation of the progression on the board provided on the right-hand page?

Can you work out the proportions of the black areas in relation to the whole square in each step of the series?

ANSWER: PAGE 105

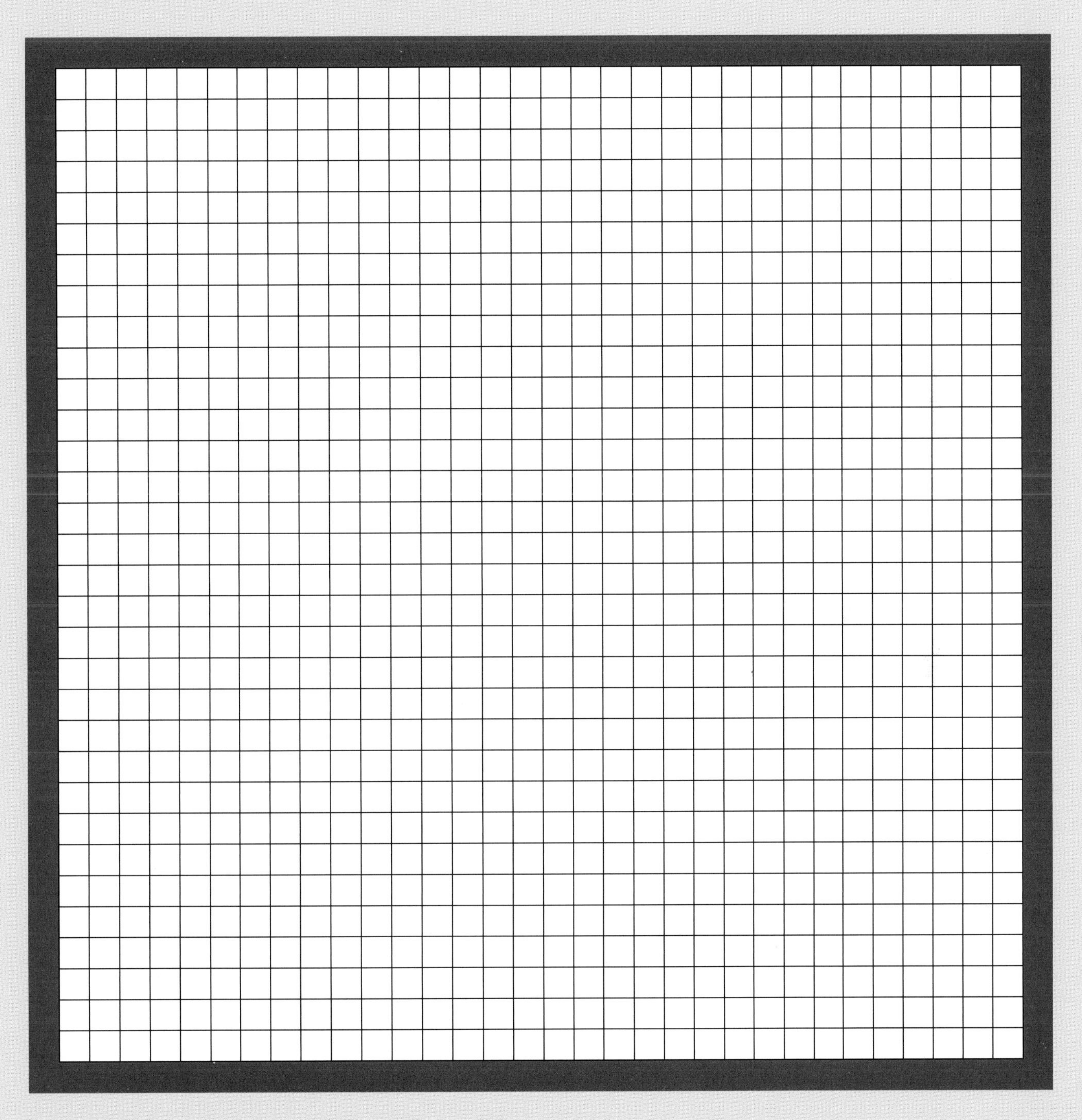

Fifth generation

This puzzle is based on the work of British biochemist Frank Odds, who devised a set of rules for generating patterns and called those patterns spirolaterals. Can you continue the sequences?

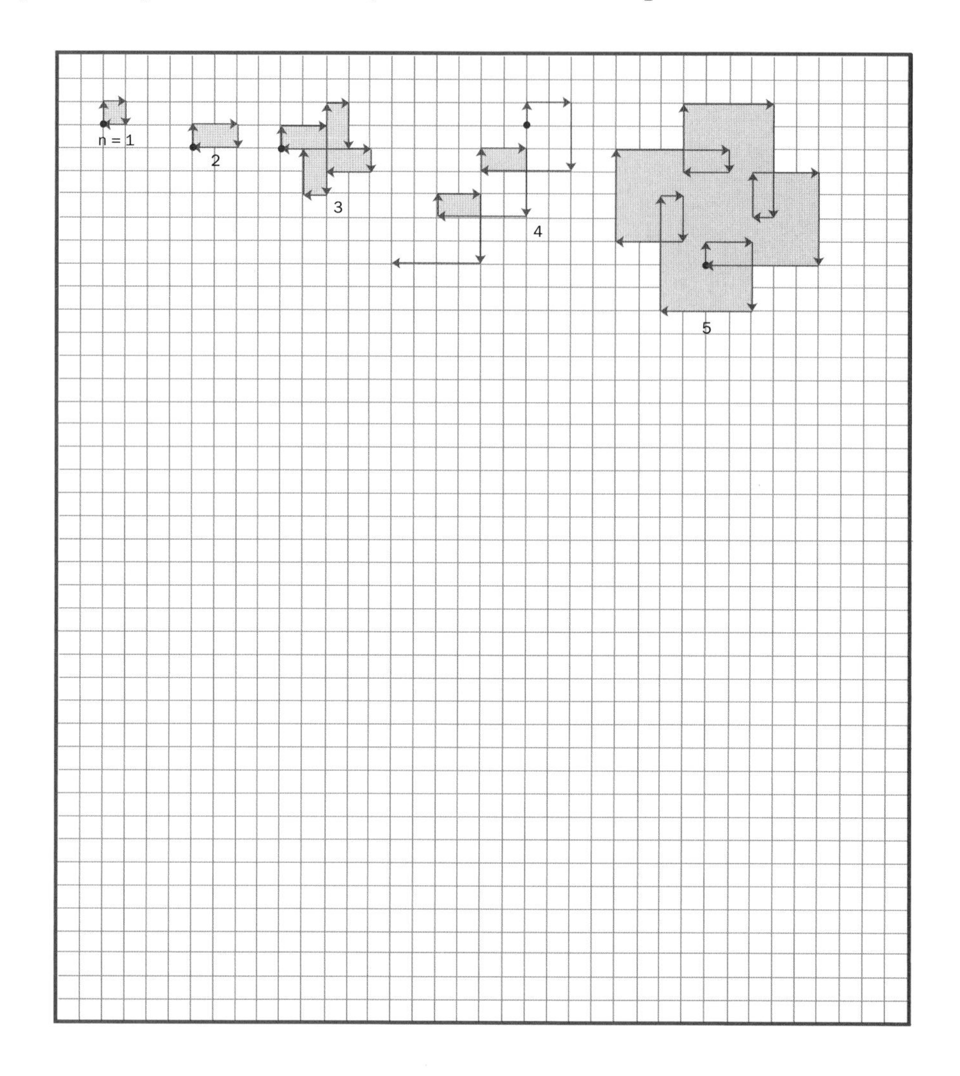

▲ SPIROLATERALS 1

Spirolaterals are geometric figures defined by the path generated by a moving point. Spirolaterals can be thought of as paths traced by a worm according to the following rules:

The worm starts moving a distance of one unit, turns, moves two units, turns, moves three units, and so on in consecutive order, always turning 90 degrees to the right, until reaching a specified limit n, which is the order of the spirolateral, and then repeating the pattern.

You can play spirolaterals as a paper-and-pencil game on a square grid.

The first five spirolaterals are shown above.

Can you continue the sequence for n = 6, 7, 8, and 9?

ANSWER: PAGE 106

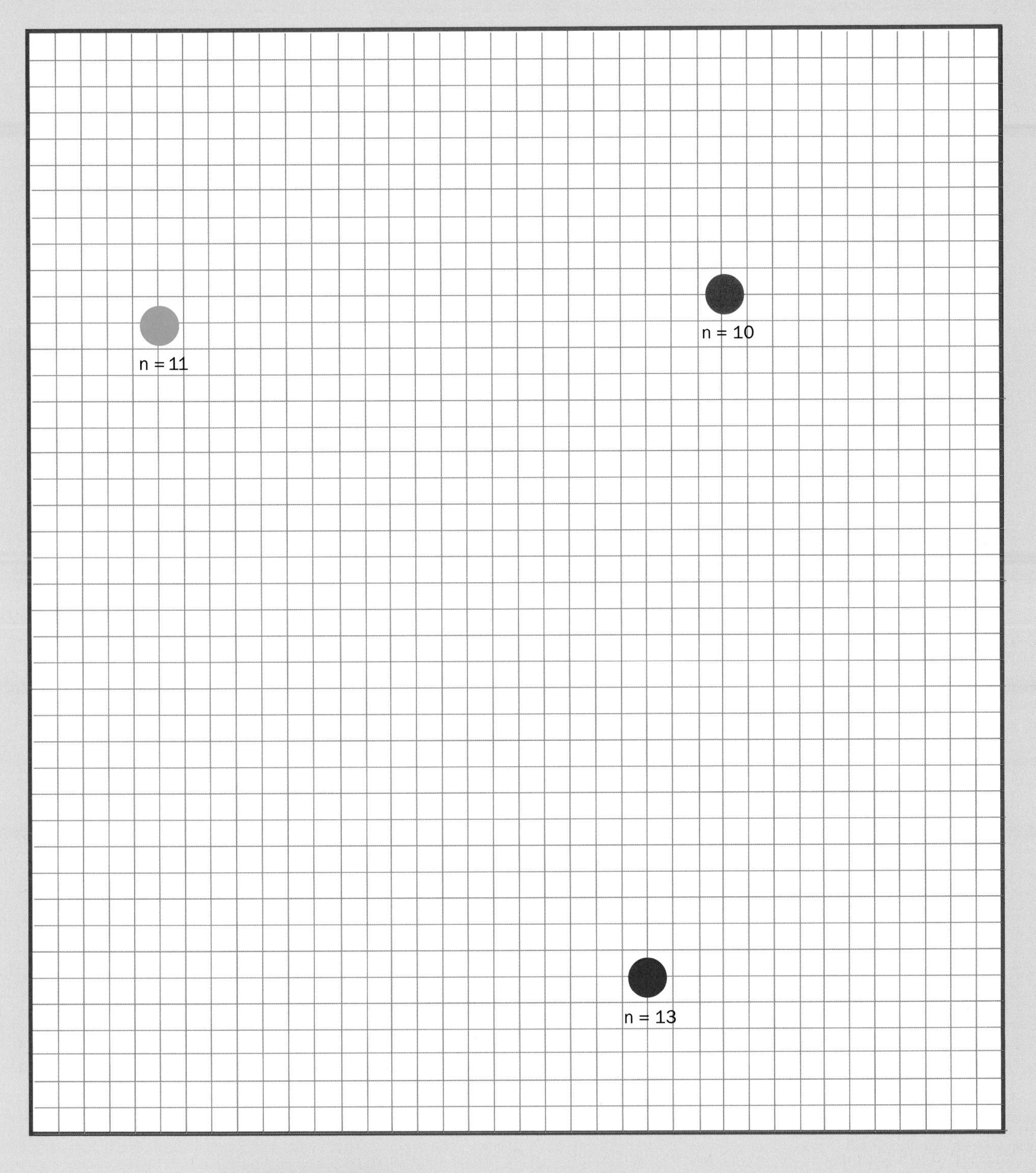

▲ SPIROLATERALS 2

Can you draw spirolaterals for $n = 10$, 11, and 13?

ANSWER: PAGE 106

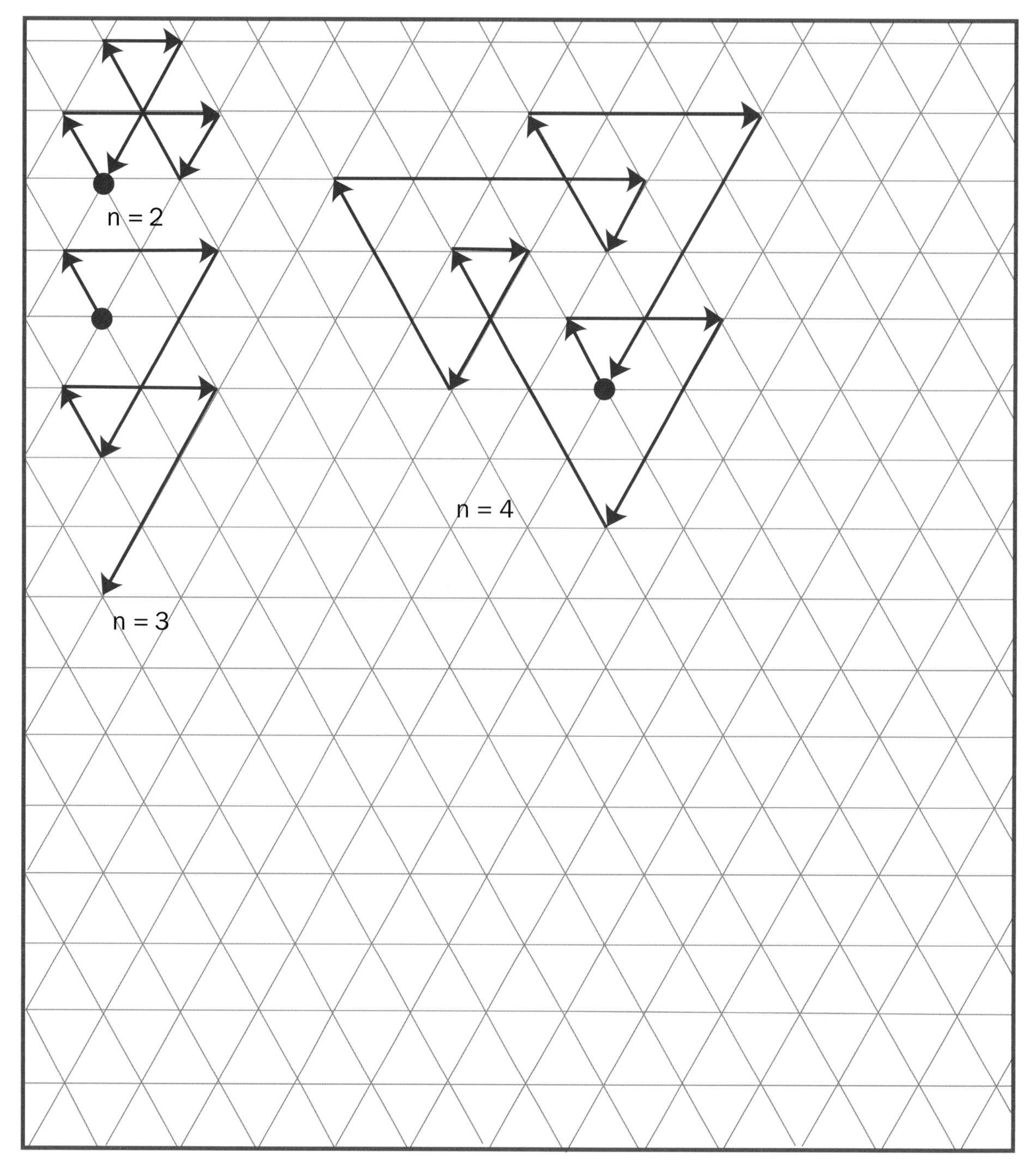

▲ SPIROLATERALS 3

Spirolaterals can have angles other than 90 degrees. Spirolaterals with 120-degree clockwise turns can be drawn on isometric paper.

Spirolaterals with 120-degree clockwise turns are shown for $n = 2$, 3, and 4. Can you draw spirolaterals with 120-degree clockwise turns for $n = 5$ and 7?

ANSWER: PAGE 106

▼ SPIROLATERALS 4

Here are some spirolaterals drawn with 60-degree counterclockwise turns. We've shown examples for n = 1, 2, 3, and 4.

Can you draw spirolaterals for n = 5, 6, 7, and 8?

ANSWER: PAGE 107

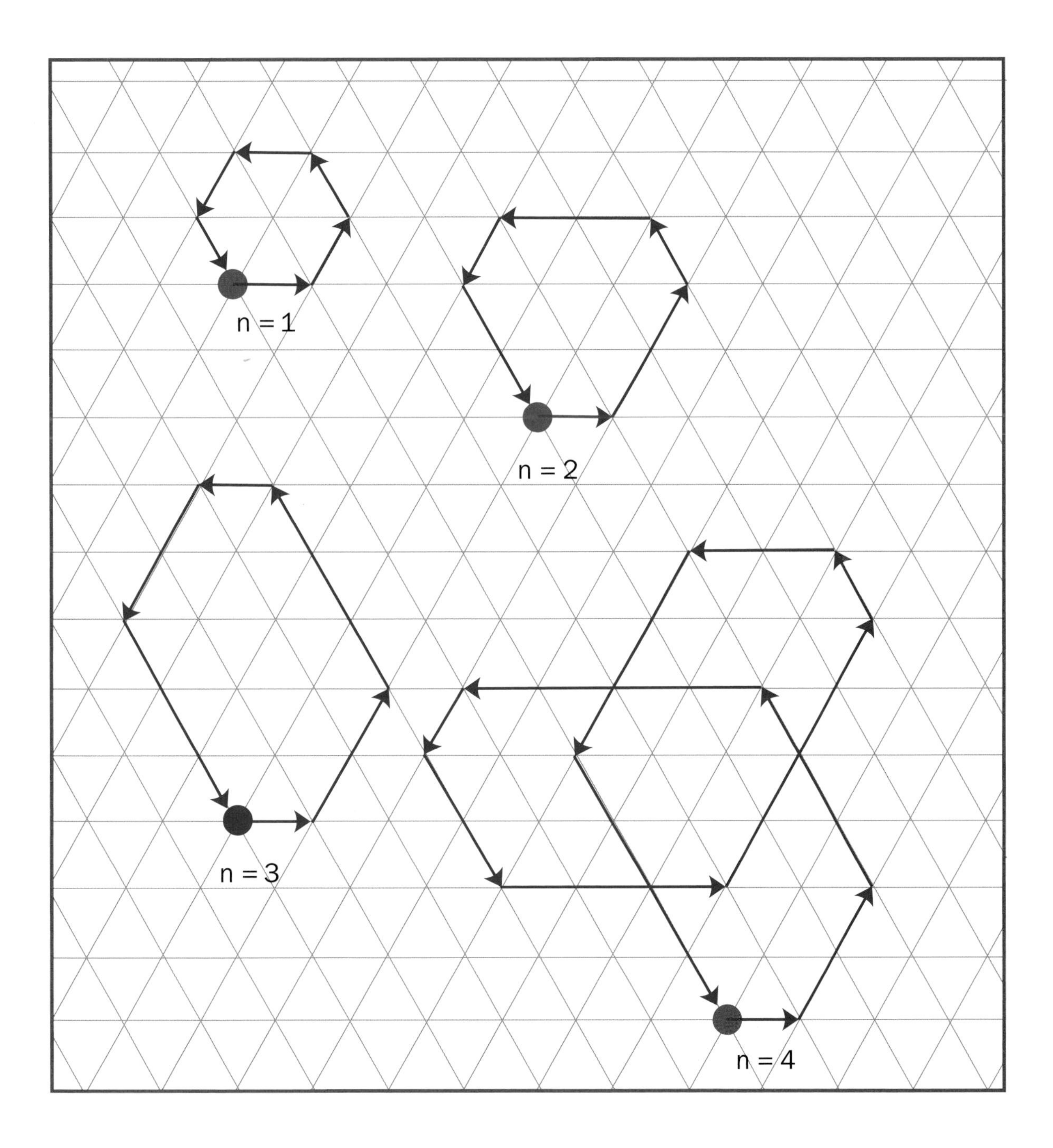

Many path games require you to find your way through a convoluted route to the finish. Can you find your way through these two puzzles, given the constraints imposed on your movements?

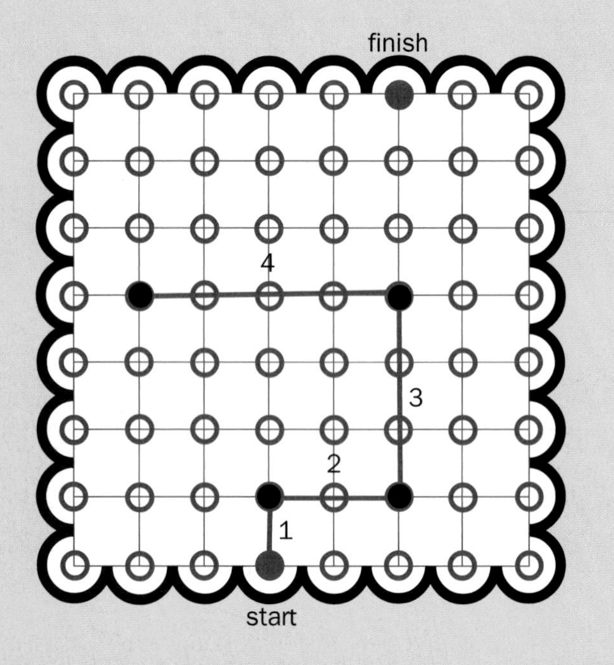

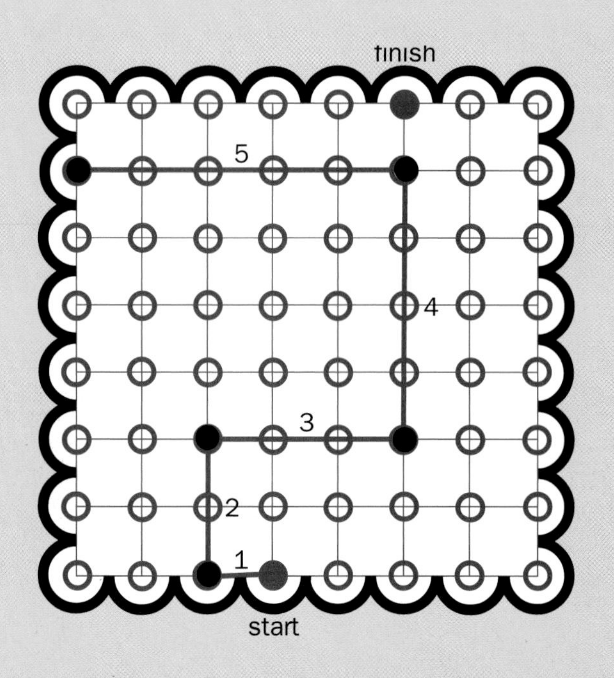

▶ LONGEST PATH

In these puzzles, the object is to reach the finish points from the start by moving in successive increasing increments—1, 2, 3, 4, 5, and so on.

Moves must be horizontal or vertical, and the path may only change directions between moves. The path may not cross itself, and the last move must stop on the finish point exactly.

Two samples are shown above, one ending after the fourth move, and another ending after the fifth—inconclusively, however.

Can you find a solution for puzzles 1 and 2?

ANSWER: PAGE 108

Puzzle 1

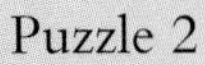
Puzzle 2

This board game for two players may appear to be very complicated at first glance. Go through the sample game and see how it works before trying it yourself.

▶ CELLULAR AUTOMATON GAME

A game for two players.

One player is blue and the other is green. The object is to end the game with the most spaces on the board covered with your color. Each space on the board will go through several stages, in this order:

Empty ▶ Yellow ▶ Red ▶ Blue or Green.

To move, a player selects a space and changes the color—and all the spaces in the same row and column—to the next color stage. So on the first move, the player will select an empty space to turn yellow, changing all the empty spaces in that row and column yellow as well.

On the next move, any yellow spaces affected by the move will turn red. After that, any red spaces will change to the color of the player making the move. Once a space has changed to blue or green, it cannot change again. The game ends when all the spaces are either blue or green. The colors are then counted to determine the winner.

A sample game finished in 13 moves with the blue player winning is shown at right.

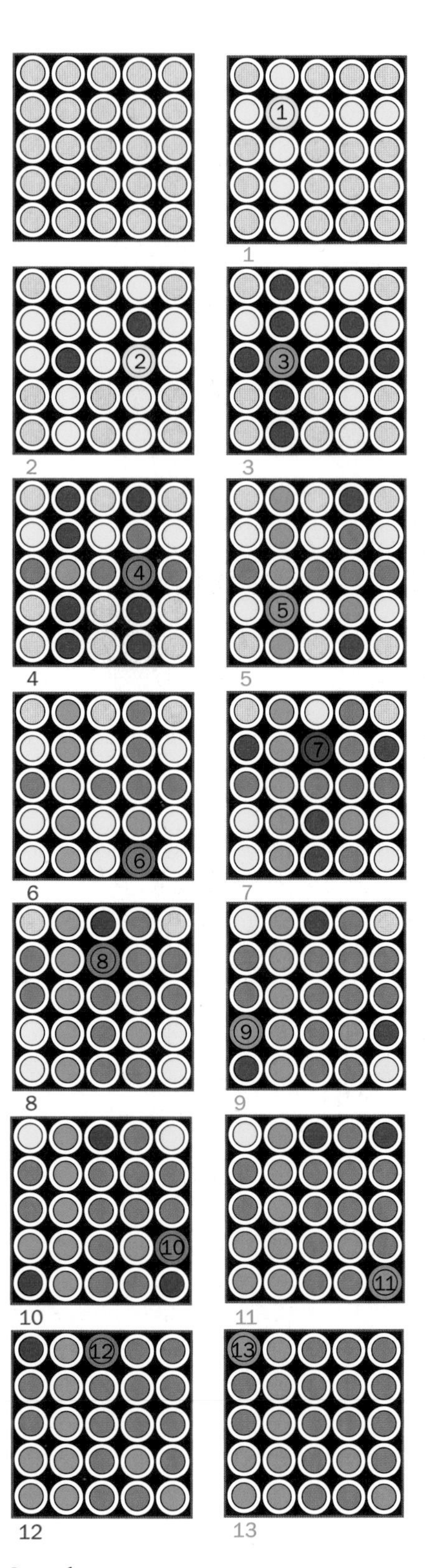

Sample game

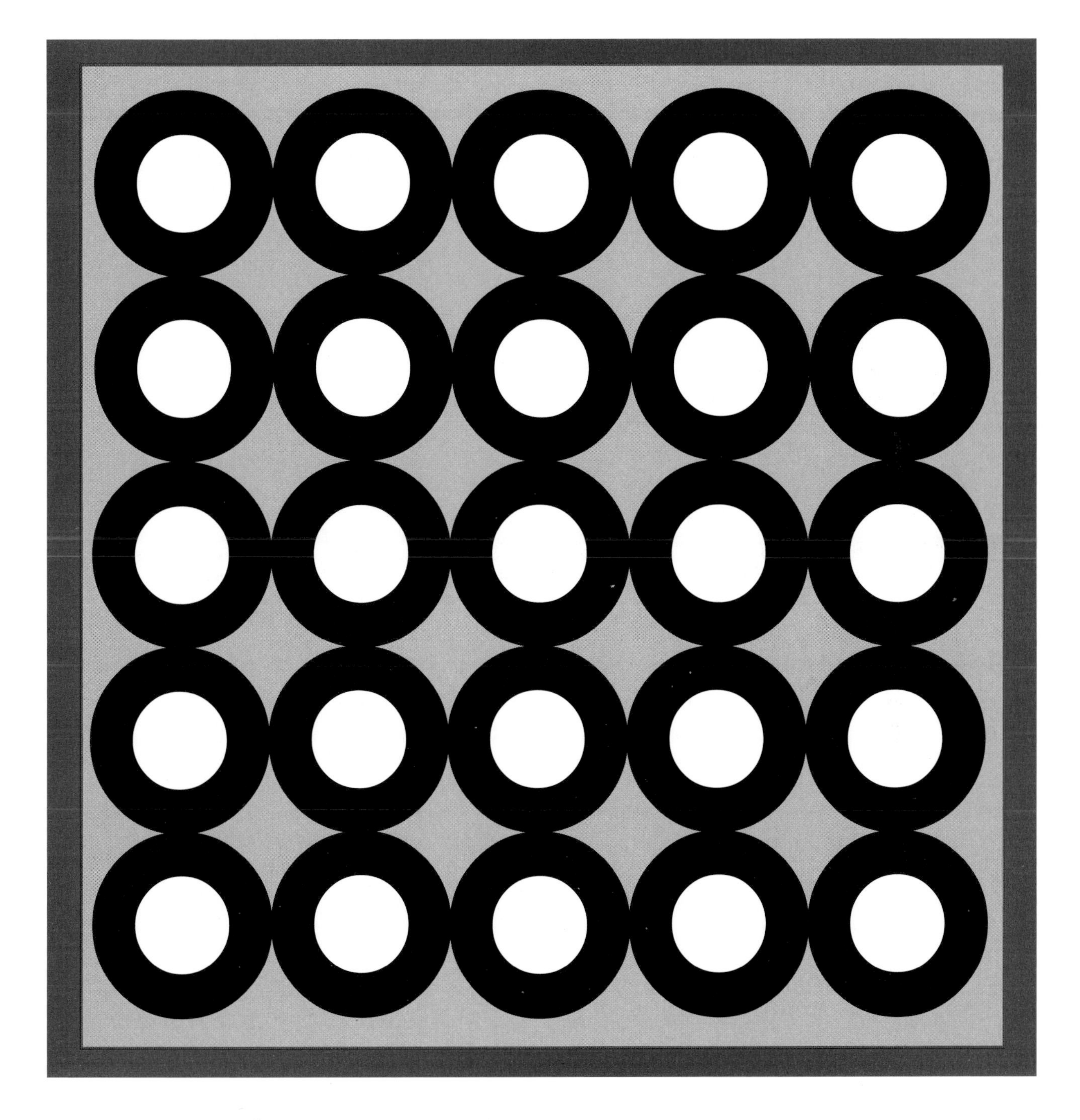

▲ CELLULAR AUTOMATON GAME BOARD

You might find it useful to copy this game board so it will be easier to get the board to lay flat. Use coins or colored poker chips to mark the appropriate colors.

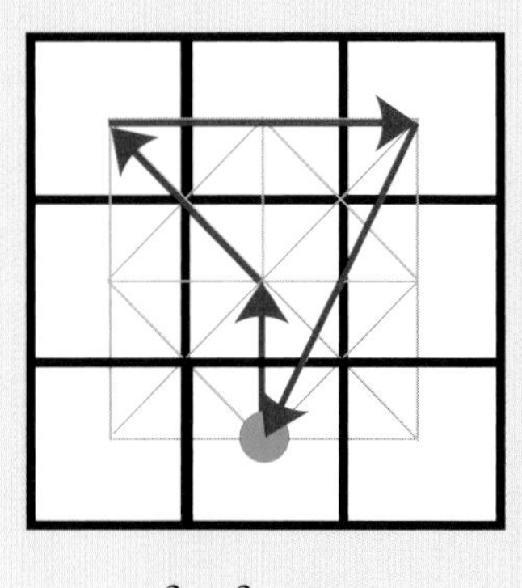

3 x 3 square
4 moves

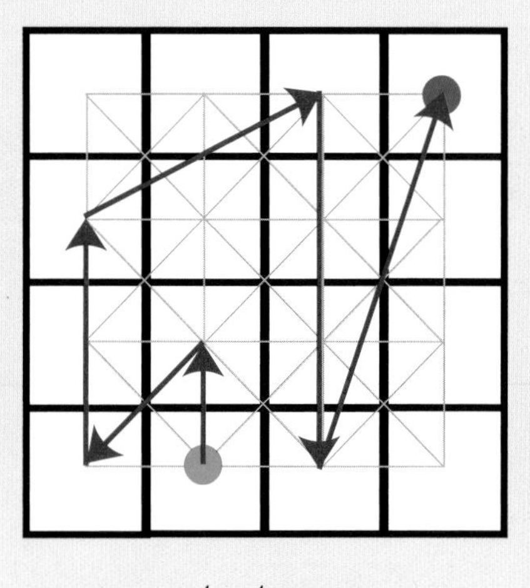

4 x 4 square
6 moves

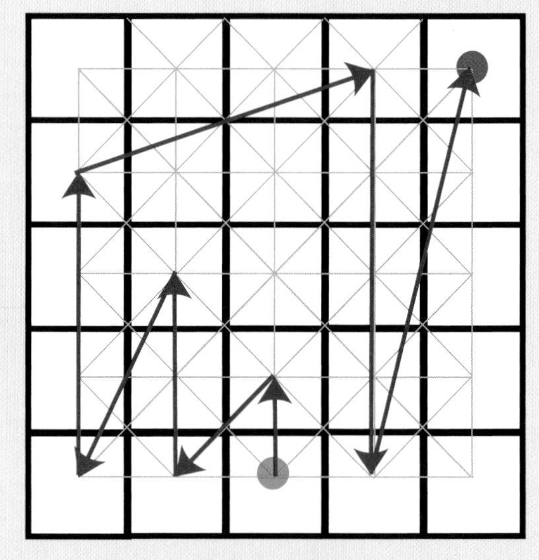

5 x 5 square
8 moves

▶ CELLULAR PATHS 1

The general object for these puzzles is to make a continuous path from the start (green circle), traveling from square to square. This is done by treating the square you start in as one end or one corner of a rectangle, and moving to the opposite end or corner. The rectangle sizes follow this pattern: 1 × 2, 2 × 2, 1 × 3, 2 × 3, 1 × 4, 2 × 4, 1 × 5, 2 × 5, 1 × 6, 2 × 6, and so on.

The maximum number of possible moves for the first four puzzles are shown. How many moves can be made on squares of sides 7 and 8?

The line may not cross itself, though it may travel to the same point more than once.

ANSWER: PAGE 109

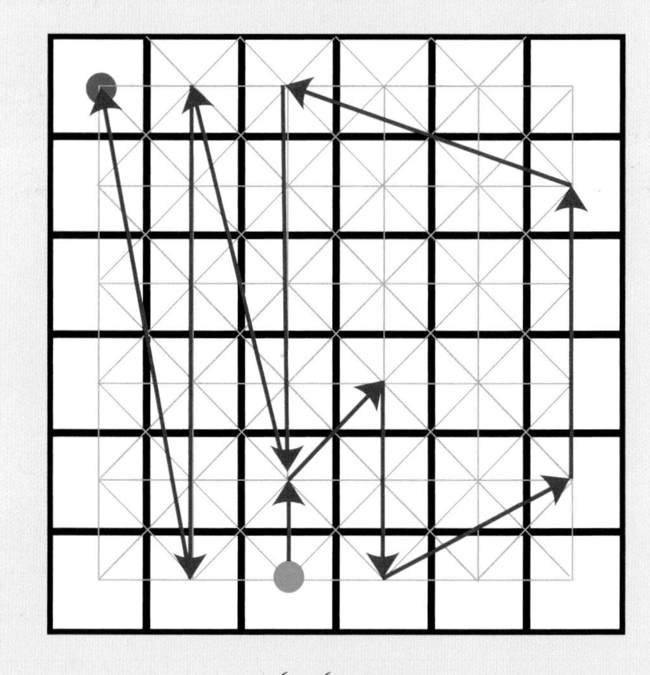

6 x 6 square
10 moves

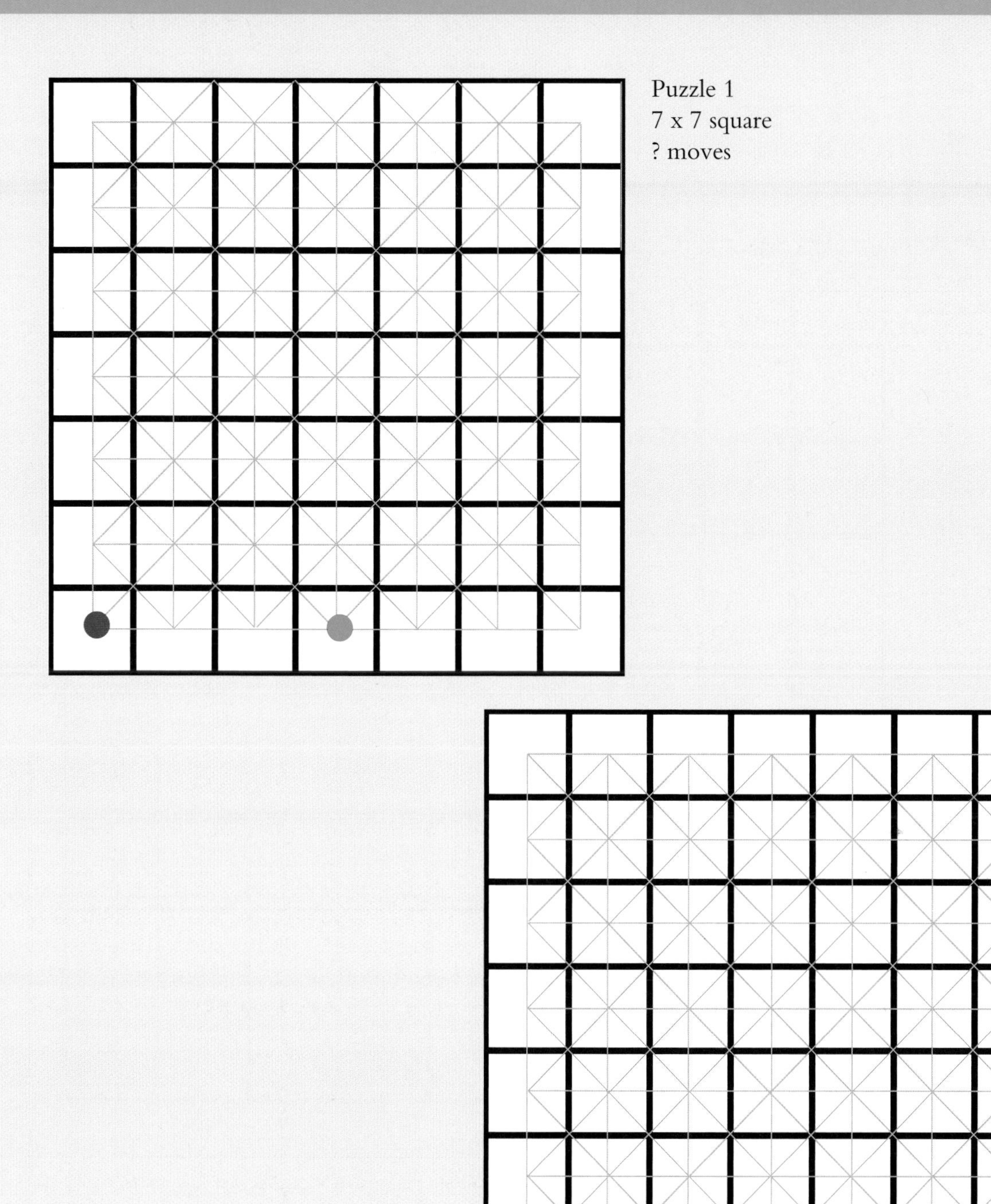

Puzzle 1
7 x 7 square
? moves

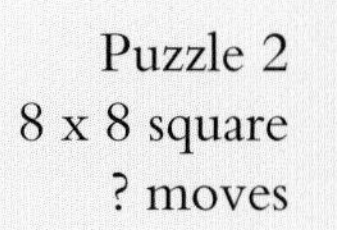

Puzzle 2
8 x 8 square
? moves

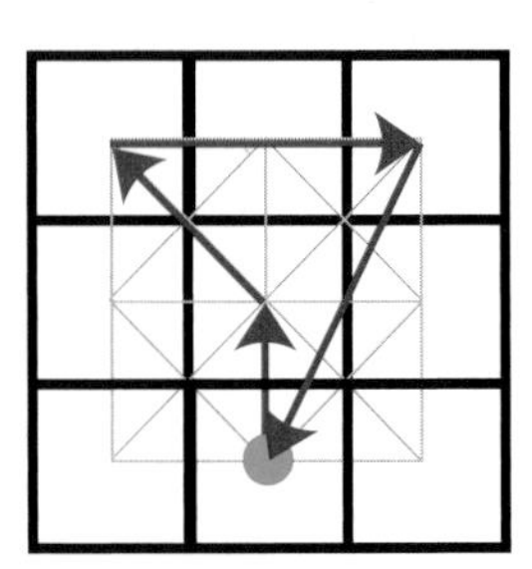

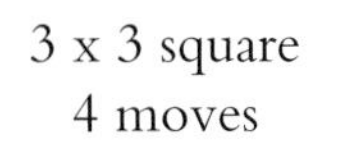

3 x 3 square
4 moves

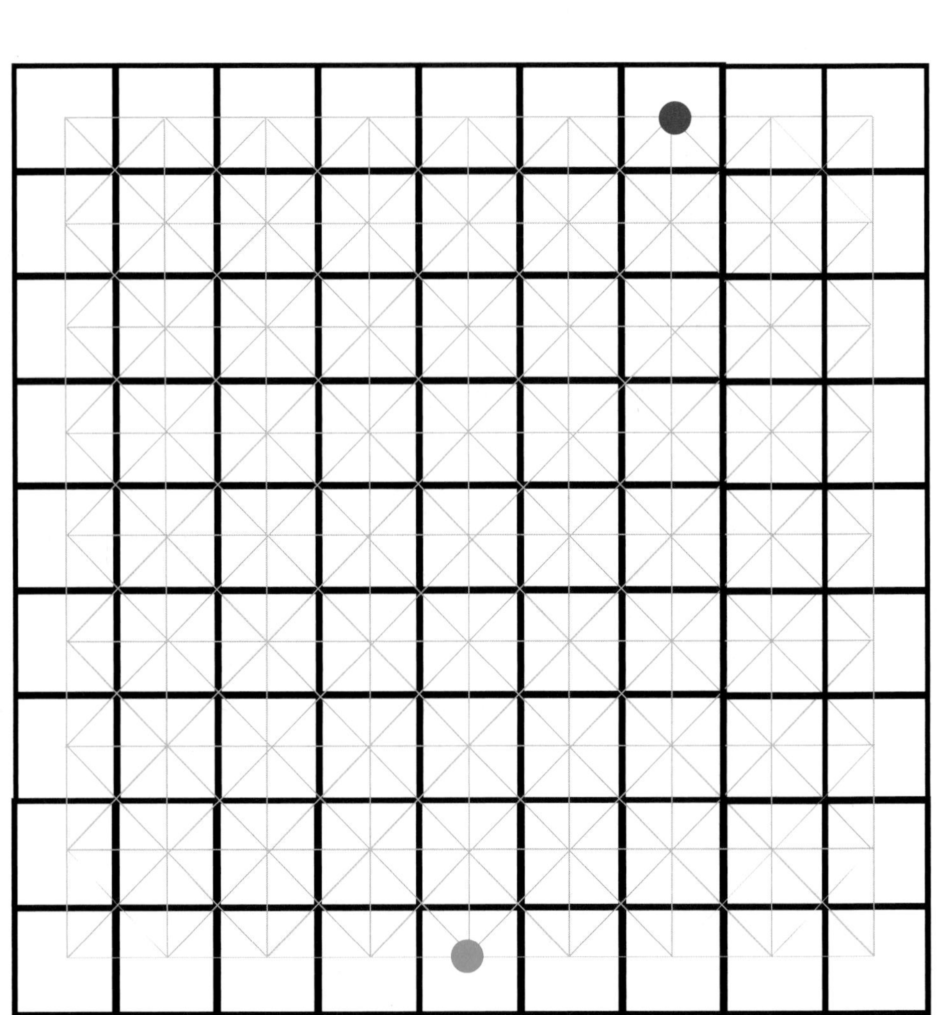

9 x 9 square
? moves

▲ CELLULAR PATHS 2

Following the same rules as the previous puzzles, can you make a continuous path from the green circle to the blue circle?

ANSWER: PAGE 109

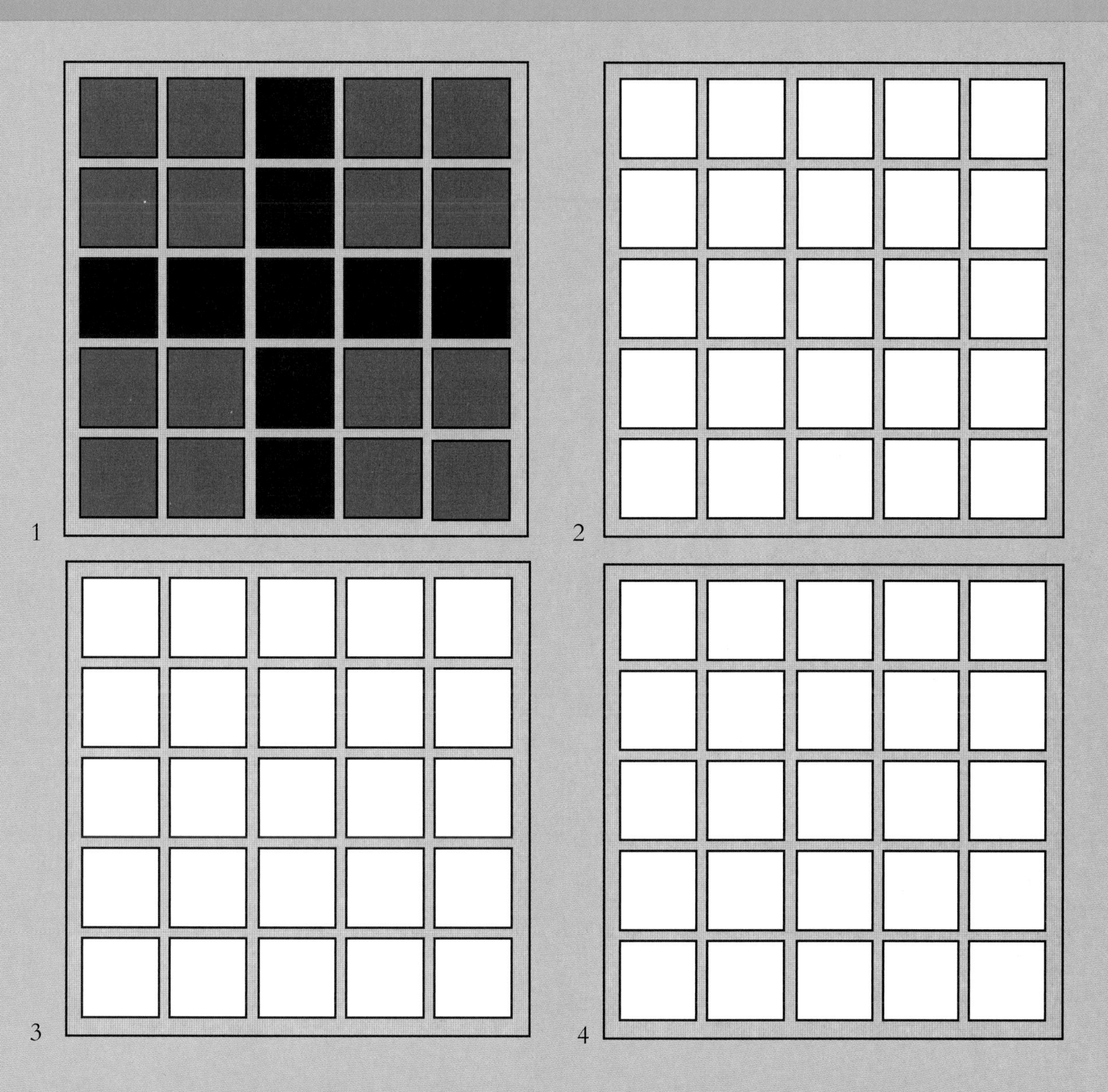

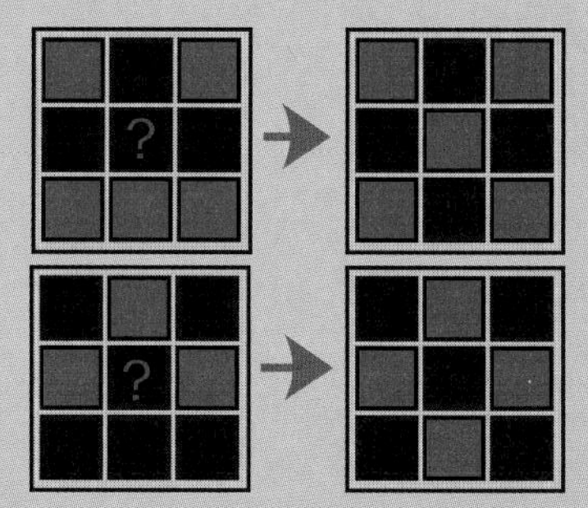

▲ CELLULAR TRANSFORMATION

The object is to transform the pattern in the initial configuration according to the following rules: In every generation, the color of each square is determined by that of the four horizontally and vertically adjacent neighbors in the generation before it.

If a black square is surrounded by more black squares than red squares, it will change color. A majority of red squares leaves it black. Red squares do the opposite: They change color if surrounded by a majority of red squares and stay red if surrounded by more black squares. For both colors, in the event of a tie, the color remains the same. See the examples at left.

How many generations are needed before all the future generations can be easily predicted?

ANSWER: PAGE 110

Square numbers, sequences, and prime numbers provide the basis for many of the puzzles found in this book. Below find two more to test your skills.

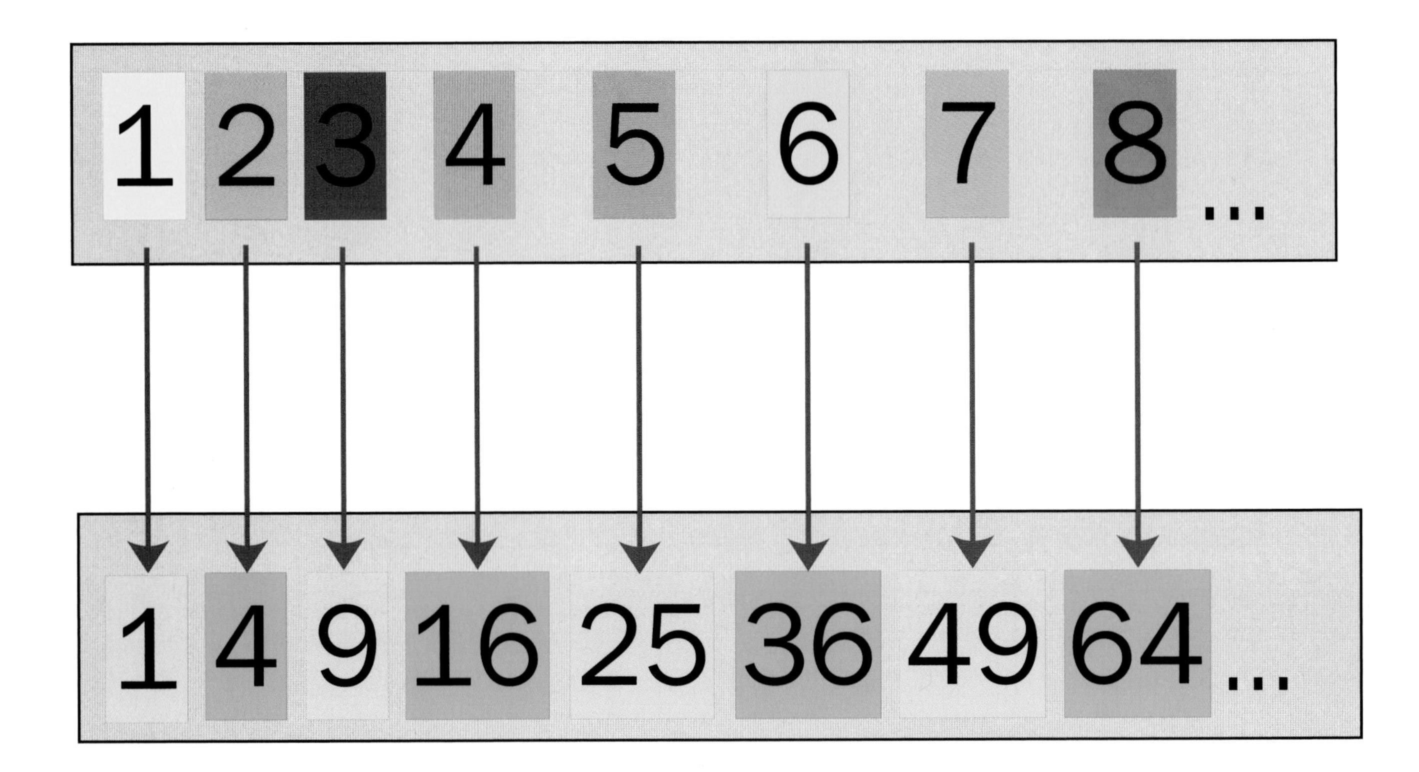

▲ GALILEO'S PARADOX

Every whole number has a square. Are there as many perfect squares as whole numbers? What do you think?

ANSWER: PAGE 111

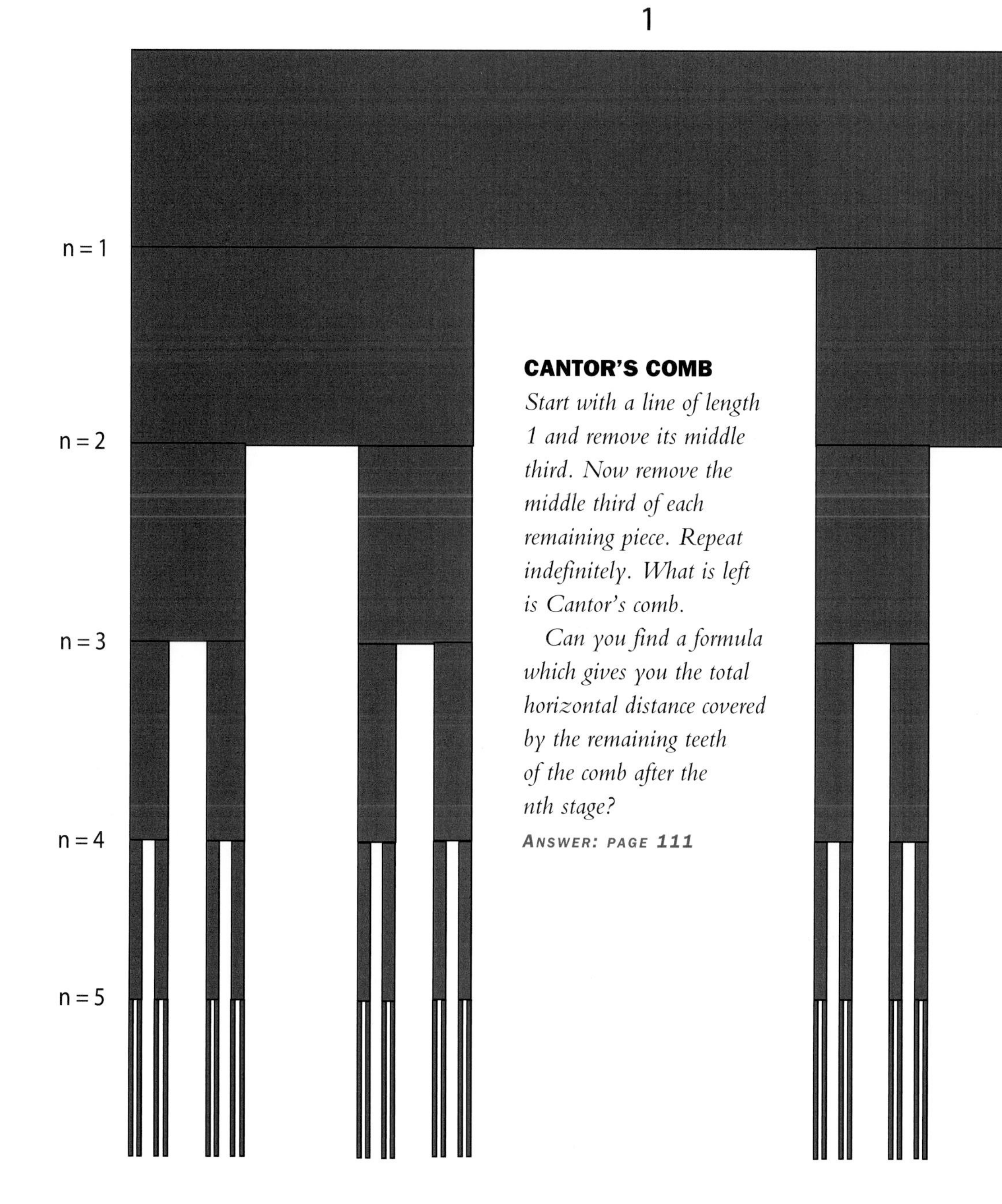

CANTOR'S COMB

Start with a line of length 1 and remove its middle third. Now remove the middle third of each remaining piece. Repeat indefinitely. What is left is Cantor's comb.

Can you find a formula which gives you the total horizontal distance covered by the remaining teeth of the comb after the nth stage?

ANSWER: PAGE **111**

▶ EIGHT ONE OUT

Can you discover the odd one out in this sequence of eight numbers?

ANSWER: PAGE 111

31

331

3331

33331

333331

3333331

33333331

333333331

▲ HOTEL INFINITY

Hotel Infinity has an infinite number of rooms. No matter how full the hotel is, there is always room for the next late arrival. The manager simply moves the person in room number 1 into room number 2, the person in room number 2 into room number 3, and so on. At the end of this process, which may take a long time, room number 1 becomes vacant for the new guest.

But how can the hotel manager cope with the problem of accommodating an infinite number of guests arriving late?

ANSWER: PAGE 111

Mathematicians through the ages have been fascinated by prime numbers. These numbers are whole numbers that can be divided only by 1 and themselves. They are like atoms—the building blocks of integers.

> ***It will be another million years, at least, before we understand the primes.***
> ***Paul Erdos, mathematician***

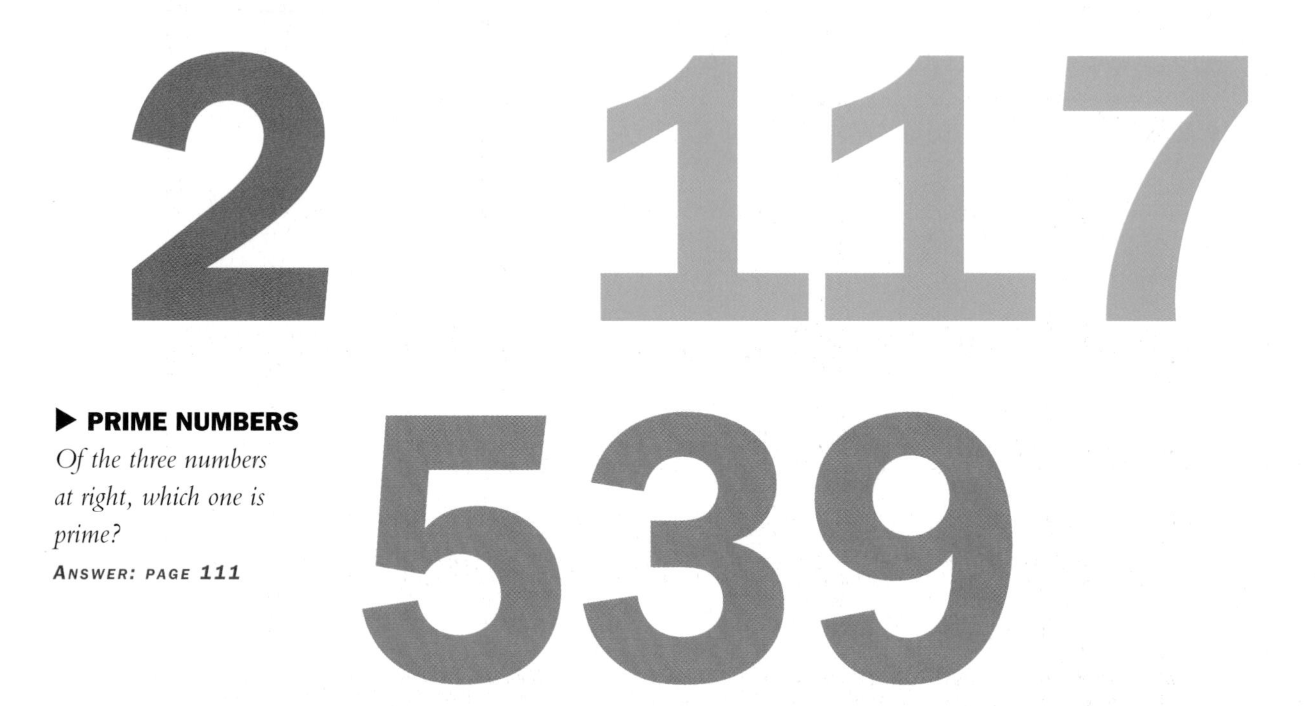

▶ PRIME NUMBERS

Of the three numbers at right, which one is prime?

ANSWER: PAGE **111**

Prime numbers

Every number is either a prime or a composite, which can be written as a product of smaller primes in exactly one way. It is believed, though not yet proven, that every even number greater than 2 can be written as the sum of two primes (Goldbach's conjecture), while the theory that every odd number greater than 9 can be written as the sum of three primes has been proven (Vinogradov's proof).

There is no formula for determining whether a number is prime or not. Even today, the sieve of Eratosthenes is the only known method of preparing a list of prime numbers up to a given limit—that is, writing down all the integers and systematically eliminating all the composite integers by first deleting the multiples of 2, then the multiples of 3, and so on. The rest will be prime numbers.

▼ SIEVE OF ERATOSTHENES

How many primes are there in the first 100 numbers?

Eratosthenes devised what he called a "sieve" for sorting out the primes from the composites. With very large numbers this method becomes difficult, but as we can see, it very rapidly provides us with, say, all the primes up to 100. Ignoring 1 (which mathematicians prefer not to call prime) take the next number in the sieve—2. That number is prime; delete all its multiples to "shake" the sieve as shown. The next number remaining (3) is also prime; delete its multiples and continue in this fashion.

How many multiples of other numbers have to be used in the sieve to obtain all the primes up to 100?

ANSWER: PAGE 112

1	2	3	4	5	6	7	8	9	10
11	12	13	14	15	16	17	18	19	20
21	22	23	24	25	26	27	28	29	30
31	32	33	34	35	36	37	38	39	40
41	42	43	44	45	46	47	48	49	50
51	52	53	54	55	56	57	58	59	60
61	62	63	64	65	66	67	68	69	70
71	72	73	74	75	76	77	78	79	80
81	82	83	84	85	86	87	88	89	90
91	92	93	94	95	96	97	98	99	100

1	11	21	31	41	51	61	71	81	91
2	12	22	32	42	52	62	72	82	92
3	13	23	33	43	53	63	73	83	93
4	14	24	34	44	54	64	74	84	94
5	15	25	35	45	55	65	75	85	95
6	16	26	36	46	56	66	76	86	96
7	17	27	37	47	57	67	77	87	97
8	18	28	38	48	58	68	78	88	98
9	19	29	39	49	59	69	79	89	99
10	20	30	40	50	60	70	80	90	100

▲ ALL NINES PARADOX

In the first ten numbers, the digit 9 appears in only one of them (10%).

Of the first 100 numbers (10^2), as we can see above, 19 numbers have a 9 in them, which is 19% or about one-fifth.

How will this proportion change for the first 1,000 numbers (10^3), shown above right? Can you guess what the percentage of numbers having a nine in them will be for a very, very large number—say, 10^{64}?

ANSWER: PAGE 113

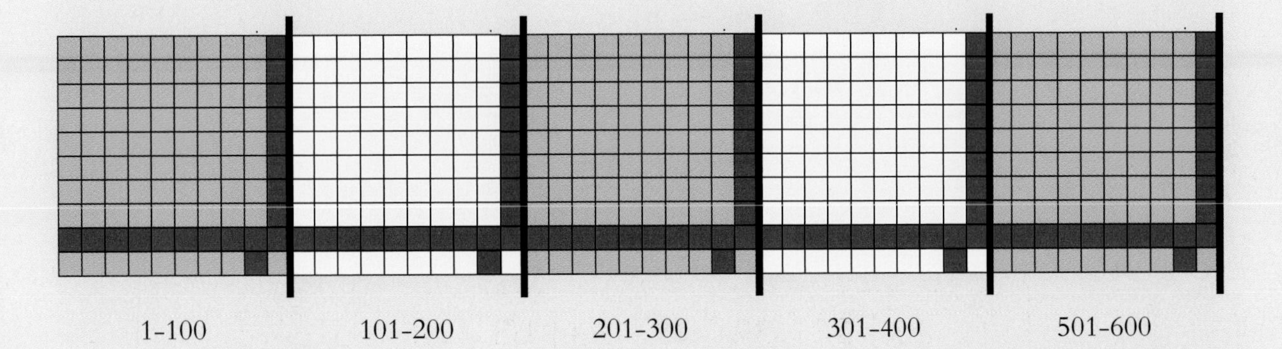
1-100
101-200
201-300
301-400
501-600

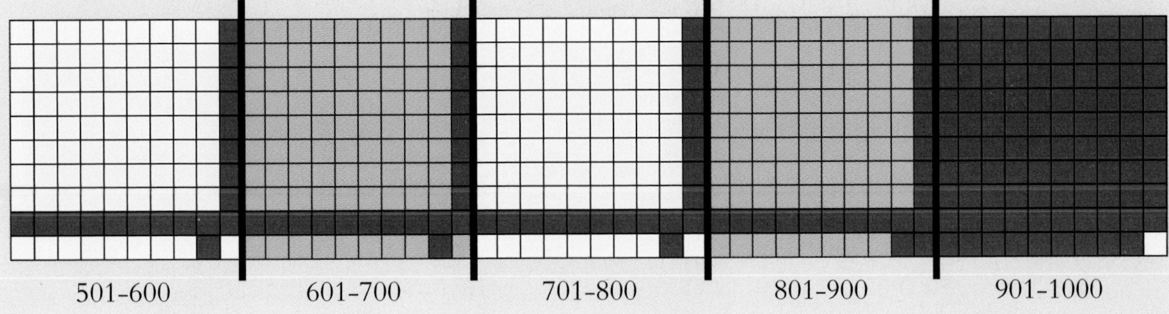

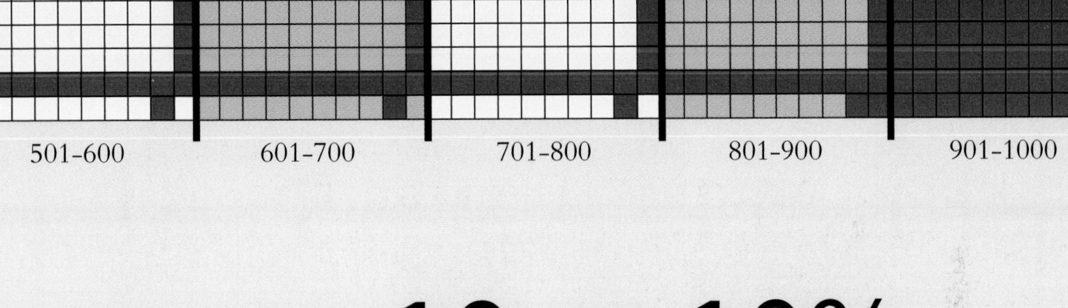
501-600
601-700
701-800
801-900
901-1000

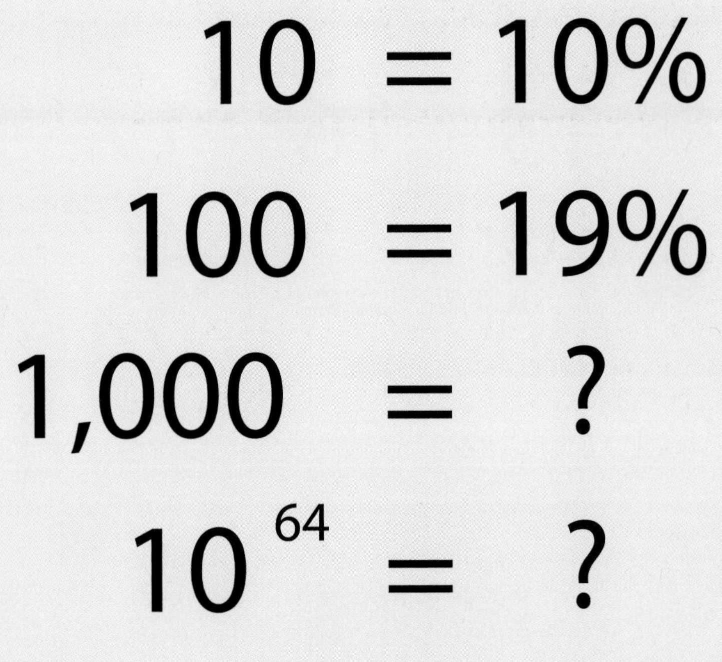
10 = 10%
100 = 19%
1,000 = ?
10 64 = ?

> ***Man has always found it easier to sacrifice his life than to learn the multiplication table."***
> ***W. Somerset Maugham***

	2	5	6	
3	4	7	8	11
10	11		15	18
12	13	16	17	20
	20	23	24	

▲ NUMBER PATTERN

Can you discover the logic of the number pattern and fit in the missing numbers?

ANSWER: PAGE 113

▼ PUNCHED CARDS PUZZLE

Copy and cut out the four square punched cards. Cut out the white holes to provide windows.

Can you overlap the four square cards so that in every small square there are four different colors visible in the circular windows?

ANSWER: PAGE 114

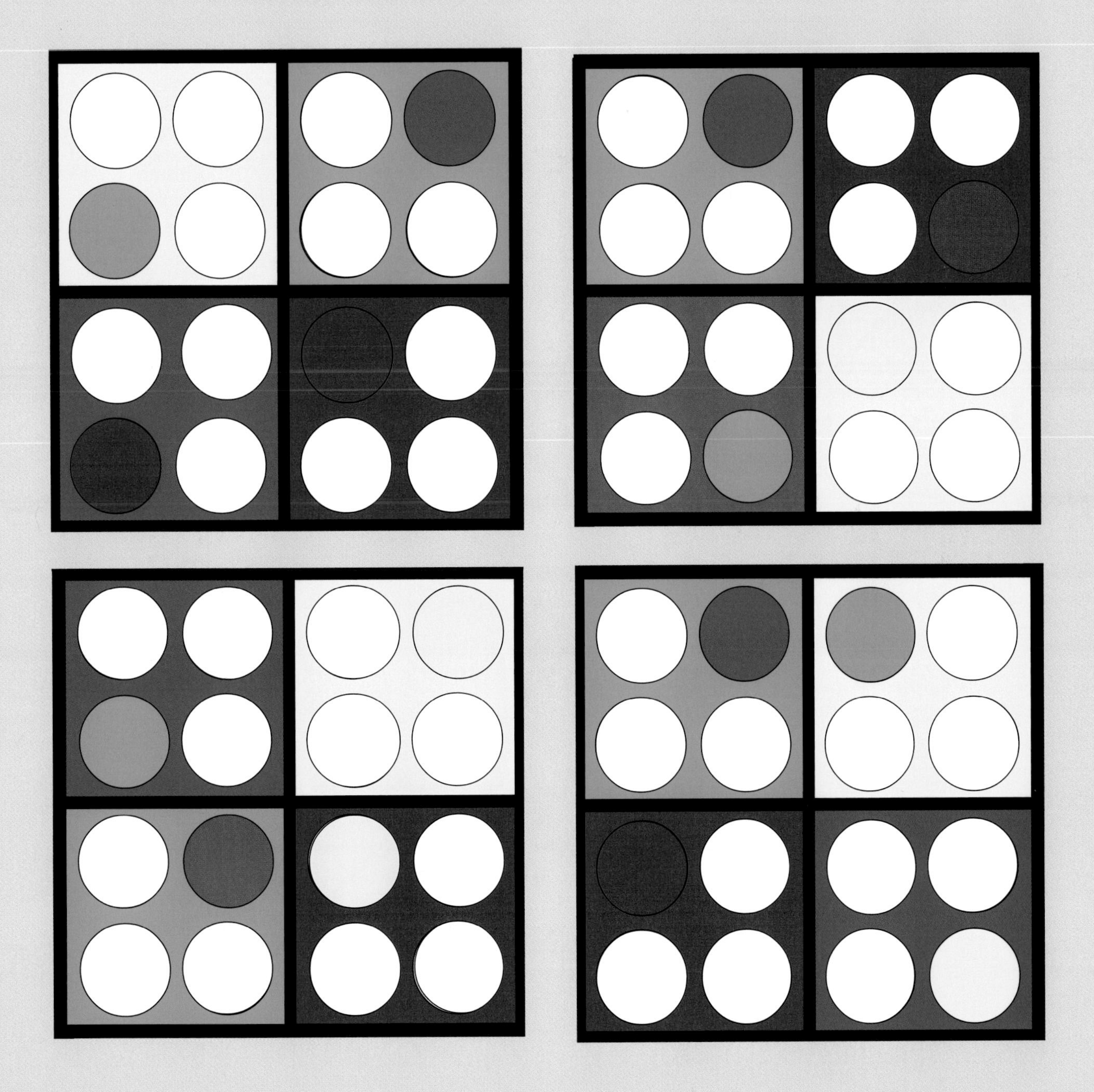

Prime spiral phenomenon

In 1963, Stanislaw Ulam, the famous Polish mathematician, was doodling numbers on a piece of paper during a boring lecture. He scribbled consecutive numbers in a square matrix, starting with 1 in the middle and spiraling outward as shown in our grid. To his utter surprise, prime numbers tended to fall on diagonal and orthogonal lines. In fact, in this matrix, the first 26 prime numbers all fall on straight lines containing at least three primes, while some diagonal lines contain even more primes.

The same mysterious line patterns appear in larger matrices as well, charting millions of primes in a spiral pattern, all forming similar configurations. A law of nature or just a coincidence? No one knows as yet. Mathematician Martin Gardner wrote in 1964, "Ulam's doodles in the twilight zone of mathematics are not to be taken lightly."

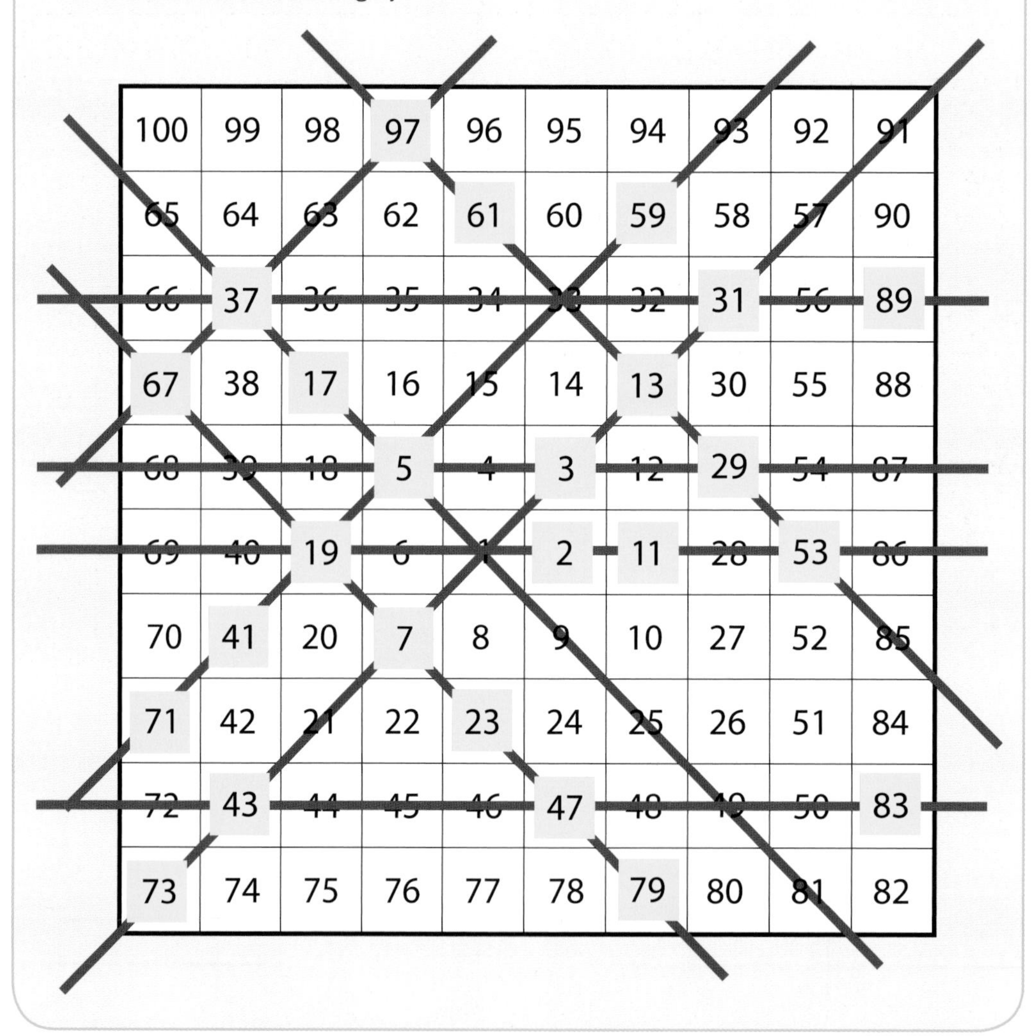

213	212	211	210	209	208	207	206	205	204	203	202	201	200	199
214	161	160	159	158	157	156	155	154	153	152	151	150	149	198
215	162	117	116	115	114	113	112	111	110	109	108	107	148	197
216	163	118	81	80	79	78	77	76	75	74	73	106	147	196
217	164	119	82	53	52	51	50	49	48	47	72	105	146	195
218	165	120	83	54	33	32	31	30	29	46	71	104	145	194
219	166	121	84	55	34	21	20	19	28	45	70	103	144	193
220	167	122	85	56	35	22	17	18	27	44	69	102	143	192
221	168	123	86	57	36	23	24	25	26	43	68	101	142	191
222	169	124	87	58	37	38	39	40	41	42	67	100	141	190
223	170	125	88	59	60	61	62	63	64	65	66	99	140	189
224	171	126	89	90	91	92	93	94	95	96	97	98	139	188
225	172	127	128	129	130	131	132	133	134	135	136	137	138	187
226	173	174	175	176	177	178	179	180	181	182	183	184	185	186
227	228	229	230	231	232	233	234	235	236	237	238	239	240	241

▲ PRIME SPIRAL

Ulam also investigated spirals that started with whole numbers other than 1, like the one shown above, which starts with 17 in the middle. He was astonished to observe the strange patterns in the distribution of prime numbers in these spirals.

Shade in the primes to see the pattern for yourself: 17, 19, 23, 29, 31, 37, 41, 43, 47, 53, 59, 61, 67, 71, 73, 79, 83, 89, 97, 101, 103, 107, 109, 113, 127, 131, 137, 139, 149, 151, 157, 163, 167, 173, 179, 181, 191, 193, 197, 199, 211, 223, 227, 229, 233, 239, 241.

ANSWER: PAGE 115

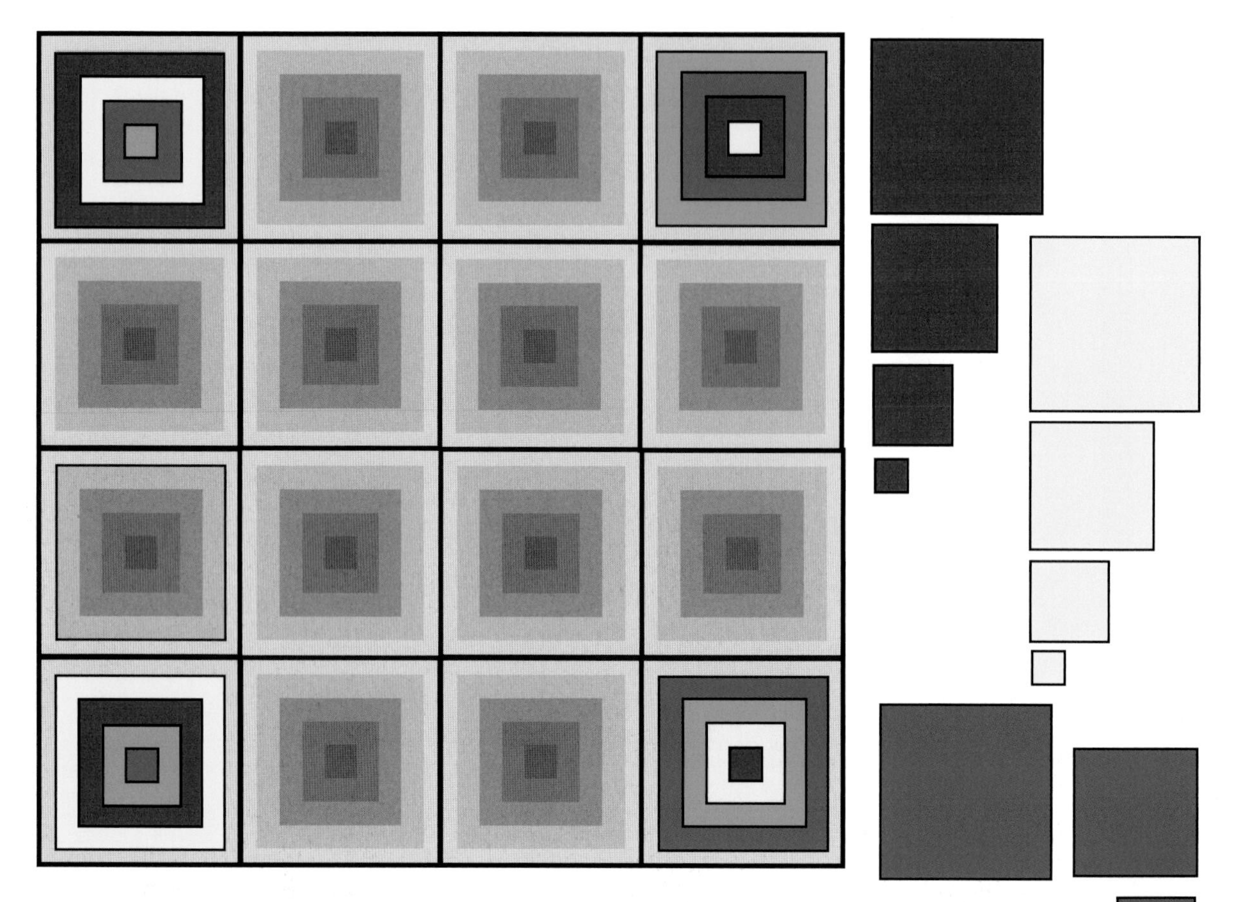

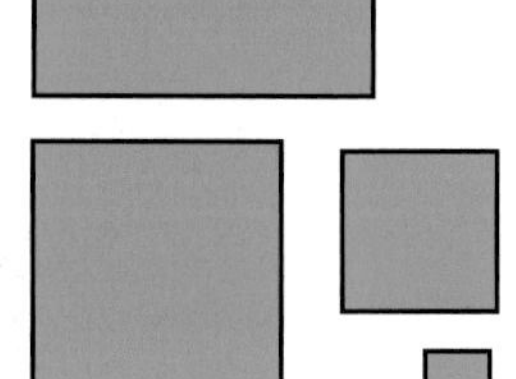

▲ TOWER POWER

In this game the 16 playing pieces are initially placed on the board as shown. Players choose their colors and alternate moves according to the following rules:

1) A player may move the top piece of a tower to an adjacent empty space in any direction or to an adjacent field occupied by a square of the same color that is one size larger.

2) A player may jump a piece over an adjacent square that is one size smaller into an empty field behind it or onto a square of the same color that is one size larger.

3) Multiple jumps are allowed; they may end on an empty field or on a square of the same color one size larger than the moving piece.

4) When jumping over a square of the same color that is one size smaller, the smaller square can be collected (placing it on top of the bigger square) forming a stage in the creation of a solid color tower.

5) A piece may never jump over a bigger square of any color.

6) Partial towers composed of two or three squares of the same color can be moved as one piece.

The winner is the player who is first to create a solid-color tower.

▼ BINARY ABACUS

The binary language is the language of computers. It is founded on the base 2 number system, and uses only 0s and 1s (represented here as white and black circles).

When digits are written in a row to represent a number, each place in the row has a different value. Different civilizations have used different number bases. Today, the most common base is 10—units, tens, hundreds, etc. We tell time in base 60.

Five numbers are written below in the binary system. Can you translate them into base 10?

ANSWER: PAGE 115

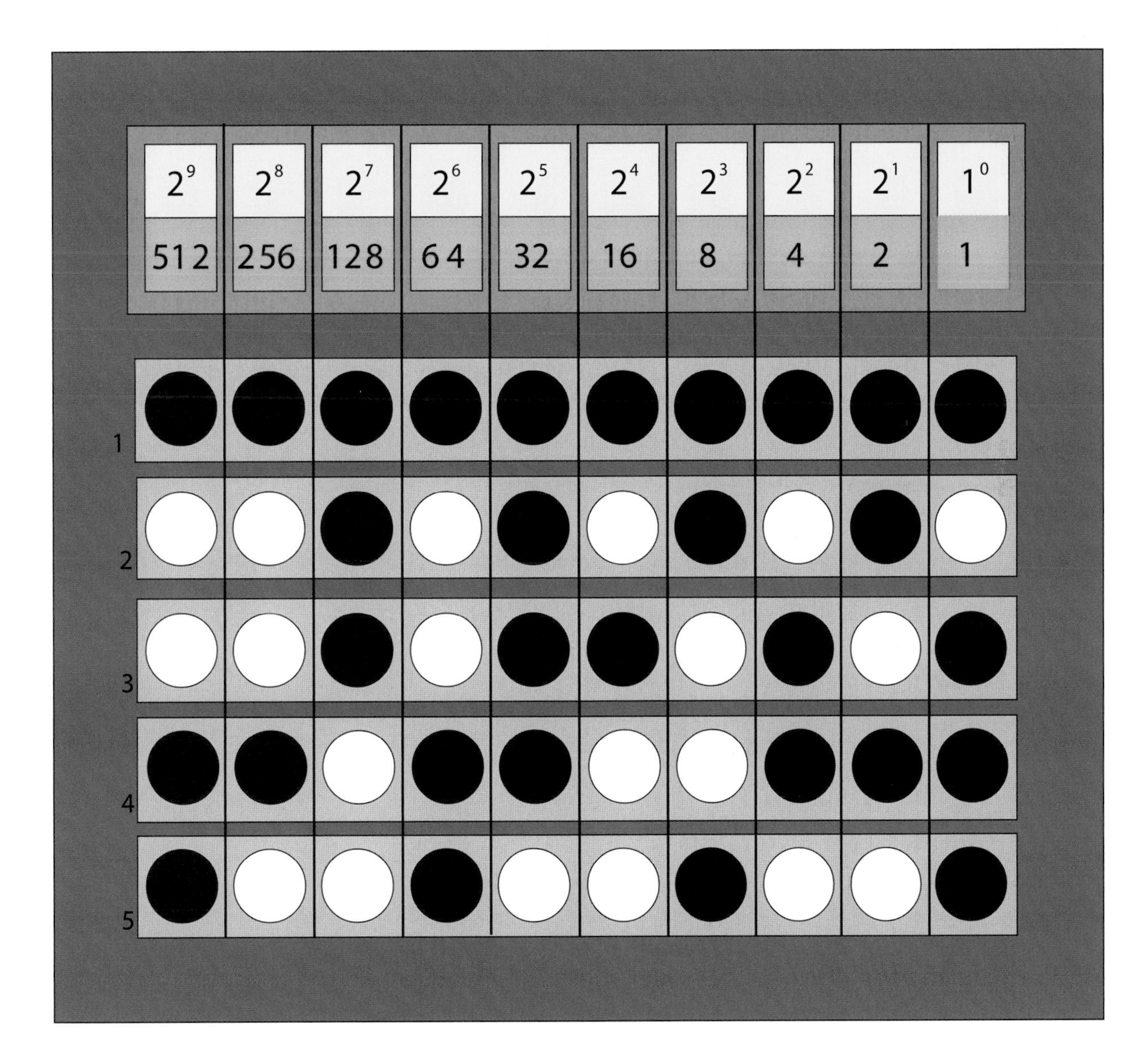

Binary wheels (like the ones below, but longer) are used to code messages in telephone transmissions and radar mapping. We are just using them for puzzles, though.

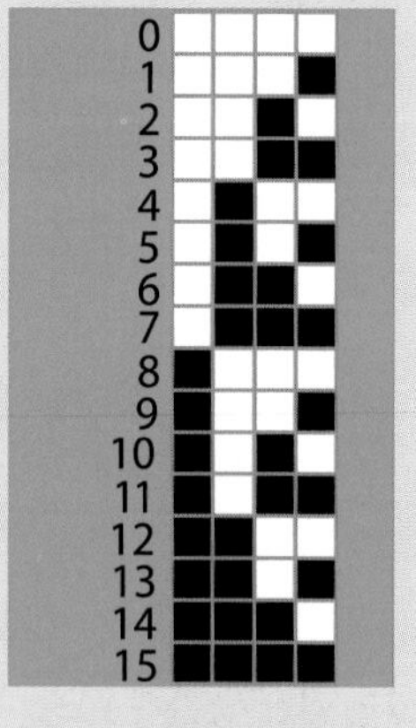

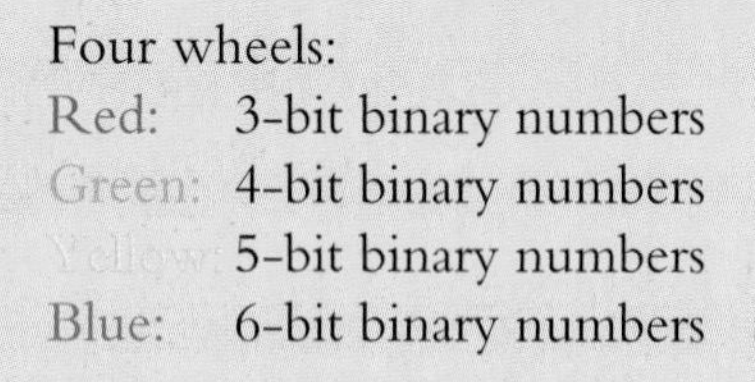
Four wheels:
Red: 3-bit binary numbers
Green: 4-bit binary numbers
Yellow: 5-bit binary numbers
Blue: 6-bit binary numbers

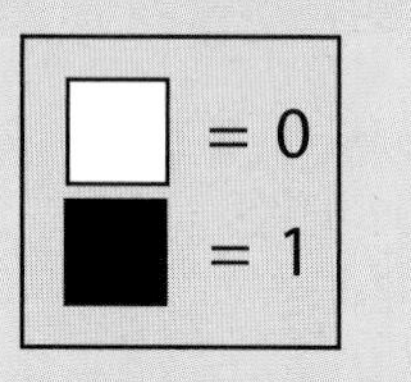

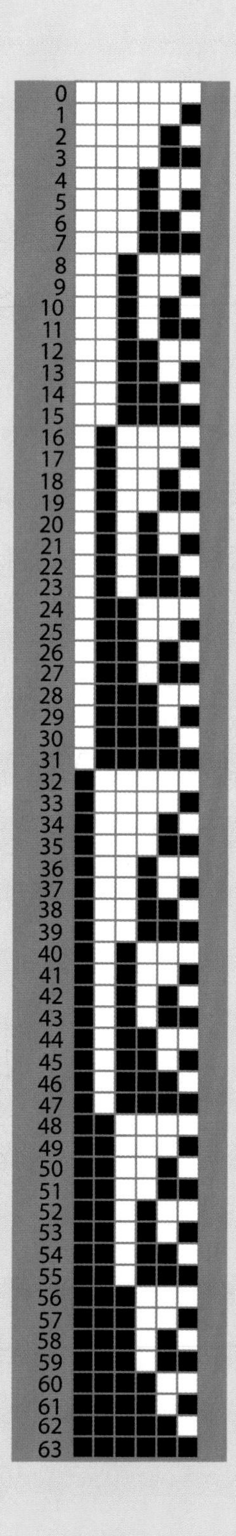

▶ BINARY WHEELS

All the possible 3-bit, 4-bit, 5-bit, and 6-bit binary numbers can be described by three, four, five, or six switches, which may be in either the "on" or "off" positions. These numbers represent the first 64 numbers (including 0) of the binary numbering system. Twenty-four switches are necessary to simultaneously express the first eight 3-bit binary numbers, 64 switches for the 4-bit binary numbers, 160 switches for the 5-bit binary numbers, and finally 384 switches for the 6-bit binary numbers.

However, in a "binary wheel," the same amount of information can be condensed to just 8, 16, 32, and 64 switches respectively—quite a difference. This can be accomplished by having the switches overlap.

Can you find a way to distribute the binary numbers along the binary wheels in such a way that all binary numbers will be represented by a set of adjacent "on" and "off" switches as you go around the wheels in a clockwise direction?

Although the switches representing each number must be consecutive, the numbers themselves need not be distributed in a consecutive sequence.

Try the 3-bit wheel first. The 4-bit wheel should be within your grasp too. Although they are difficult, we have provided the solutions for the 5-bit and 6-bit wheel as well, if you really want a mental workout!

ANSWER: PAGE 115

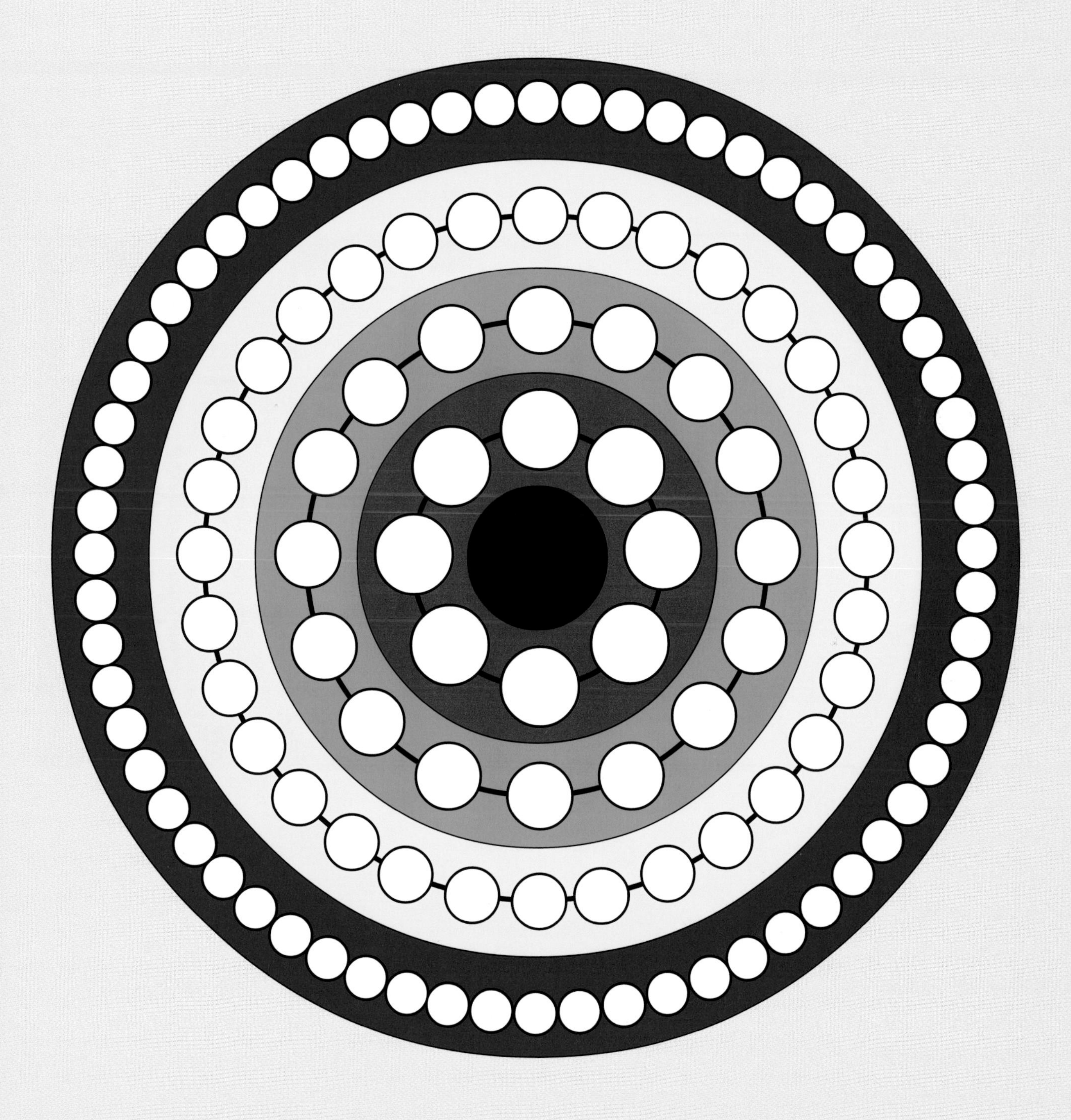

This game for one or more players will help you become familiar with the binary language of the computer. Try it and see how many solutions you can come up with.

▶ BINARY-BASED TILES

Each of the thousands of electronic circuits in a computer can switch on and off incredibly fast. When a pulse of electricity flows through a circuit we say it is on. When no electricity flows through it, we say that it is off.

The digits 1 and 0 are the basis for the binary system used in computers. Each number is called a bit—short for "binary digit." "On" circuits have a value of 1 (red); "off" circuits have a value of 0 (yellow).

Let's say that we have a set of four switches in a 2-by-2 square configuration. We have used the template at right to draw up playing tiles based on the 16 different positions that can be created from the switches.

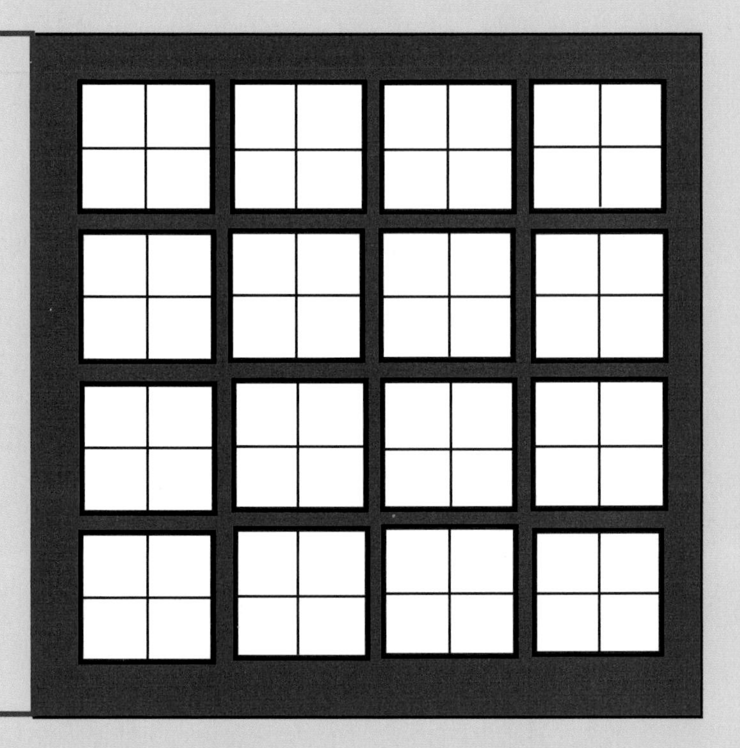

▶ Q-BITS GAME

The game for 2 players: The tiles are mixed facing down. In turn, each player takes a tile and places it on the board. It must fit exactly over a square on the board. You are allowed to rotate the tiles before placing them on the board. If a tile touches any other tile or tiles along an edge, the colors must match. The first player unable to place a tile on his turn is the loser.

The longest game can last for 16 moves. What is the shortest possible game (that is, what is the smallest number of tiles that can be placed on the board so as to make any further placing of tiles impossible)?

ANSWER: PAGE 116

Solitaire puzzles: There are many different ways to arrange the 16 playing tiles which you will find placed around the 4-by-4 game board opposite. But is it possible to do this in such a way that the colors of adjacent tiles match everywhere, domino-style? How many different solutions can you find? Use the grid on page 64 to find and record as many different solutions as you can out of the total of 50 possible.

▼ Q-BITS GAME

Game board and the set of 16 playing pieces.

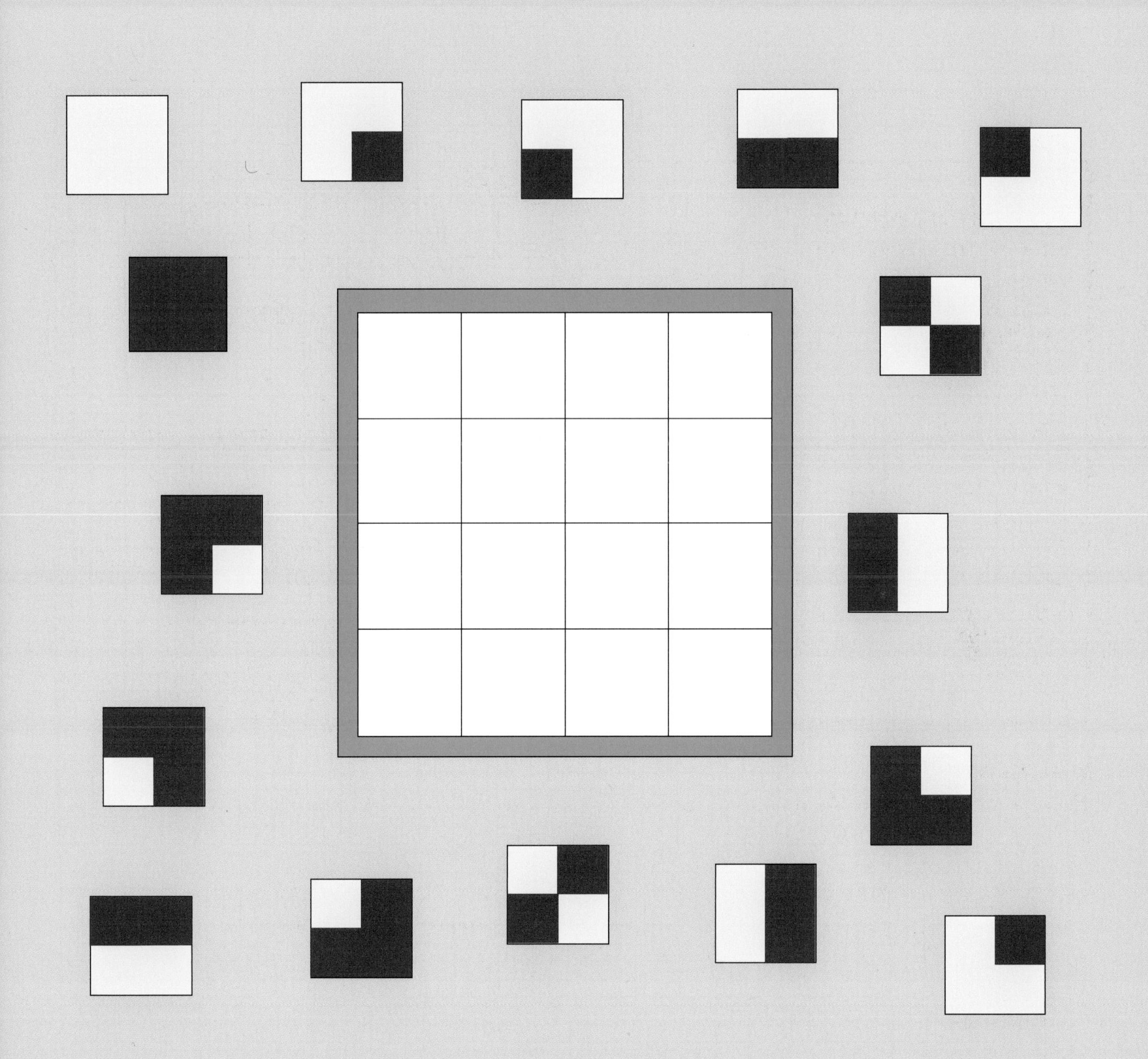

▼ Q-BITS GRIDS

Grids for recording and preserving solutions from the solitaire puzzles. There are 50 distinct different solutions. One of them is shown. How many can you find?

ANSWER: PAGE 116

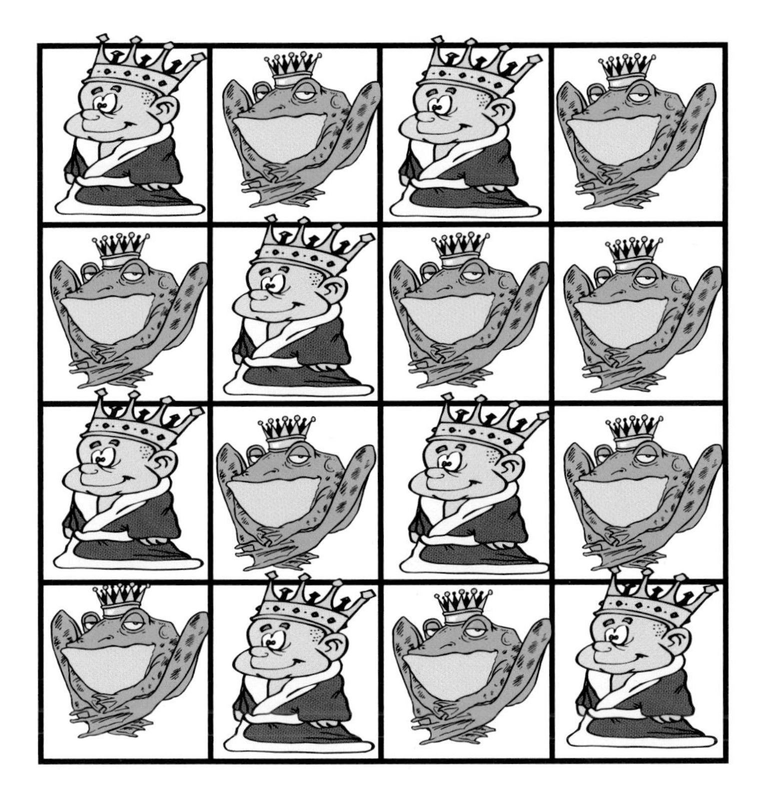

▶ FROGS AND PRINCES

Sixteen tiles are randomly distributed on a 4-by-4 board. On one side of each tile is a frog and on the other side is a prince.

The object of the game is to flip over the tiles until they all show either frogs or princes.

The tiles are flipped according to one simple rule: On any turn, you must flip all the tiles in any single row, column, or diagonal. (Note that the diagonals may be short—even a corner counts as a one-tile diagonal.)

Two random starting configurations are shown. Can you tell whether they can be solved? Is there a quick way to determine whether any configuration is solvable?

ANSWER: PAGE **117**

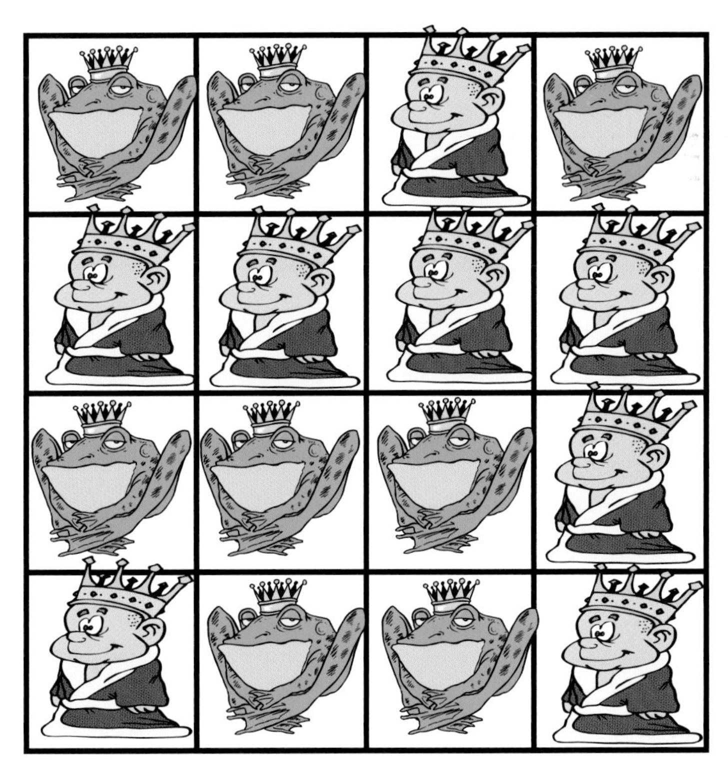

In this context "bottoms up" has nothing to do with drinks and all to do with using your head. Can you work out this glass conundrum and the one on the opposite page?

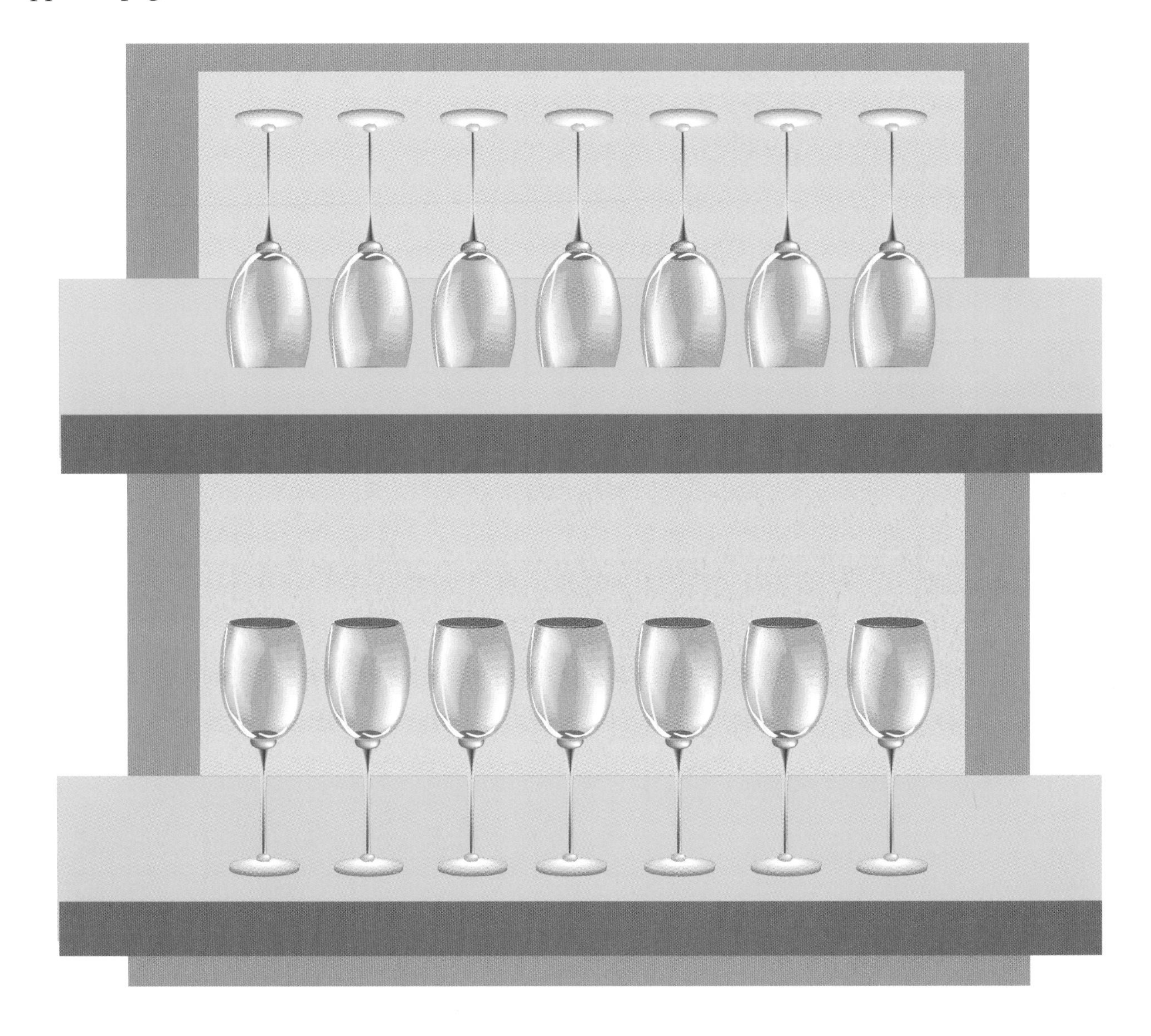

▲ GLASSES UP

The object is to turn all seven glasses face up by inverting three glasses in each move.

In how many moves can it be done?

Answer: page **117**

▲ TEN GLASSES PROBLEM

Ten glasses are placed as shown, five facing up, and five facing down. Take any two glasses and reverse both. Continue reversing the pairs of glasses as long as you please.

Can you end up with all the glasses upright?

ANSWER: PAGE 117

Much like a sliding puzzle, these binary puzzles require that you only move the patterns vertically and horizontally in order to reach the final configuration.

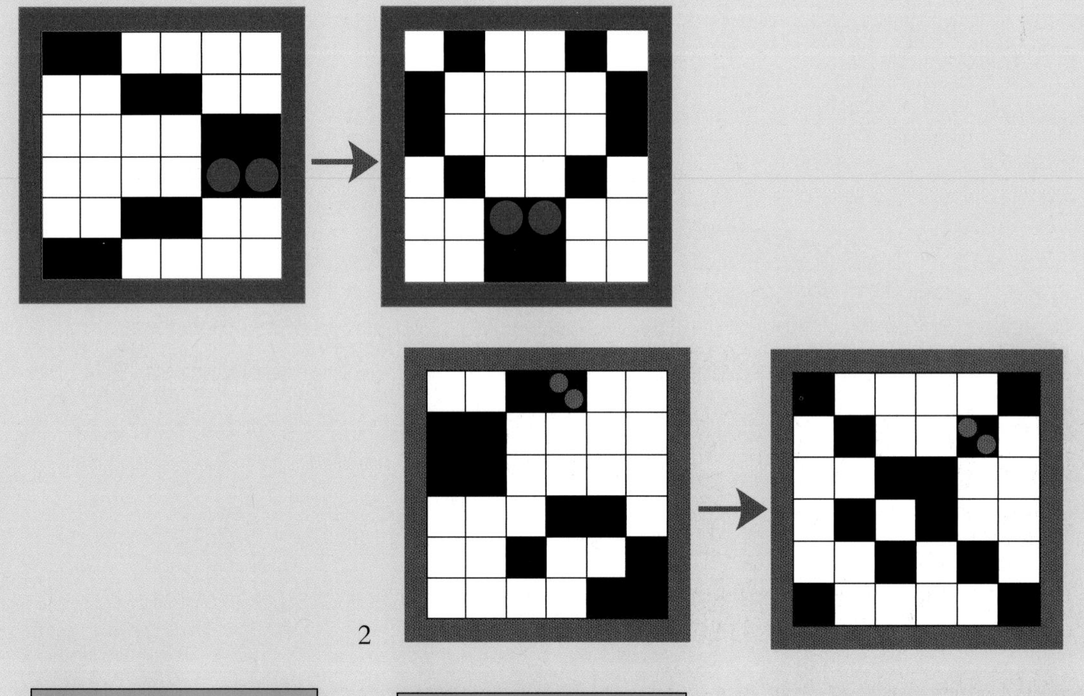

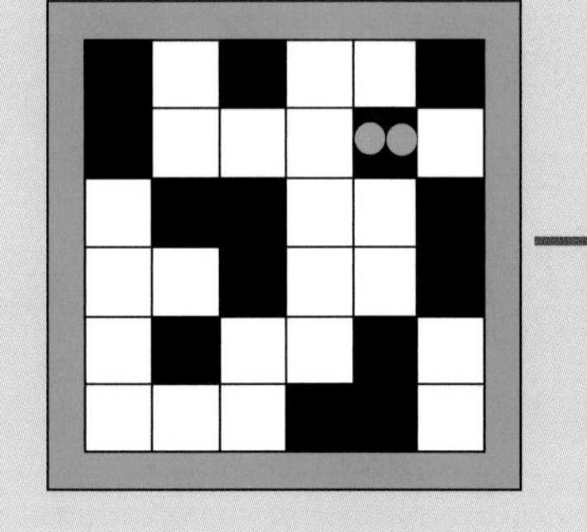

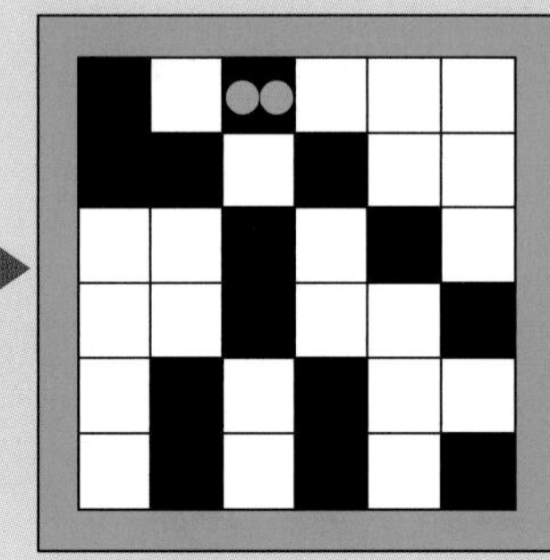

3

BINARY TRANSFORMATIONS

In these nine puzzles, the object is to transform the first pattern into the second one by swapping entire horizontal and vertical rows.

Can you find a systematic way of exchanging pairs of rows and columns to achieve each transformation?

ANSWER: PAGES 118–9

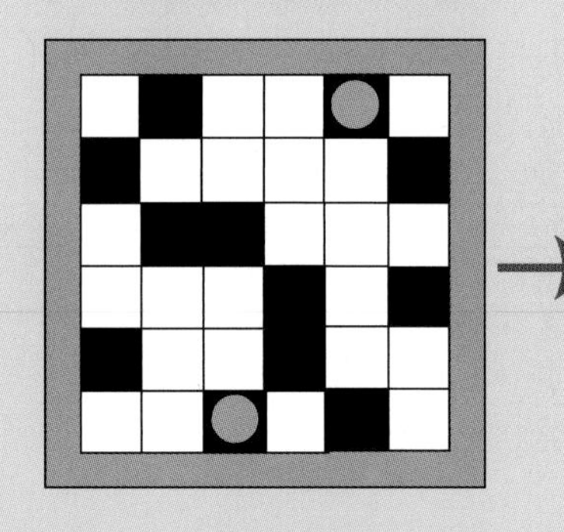

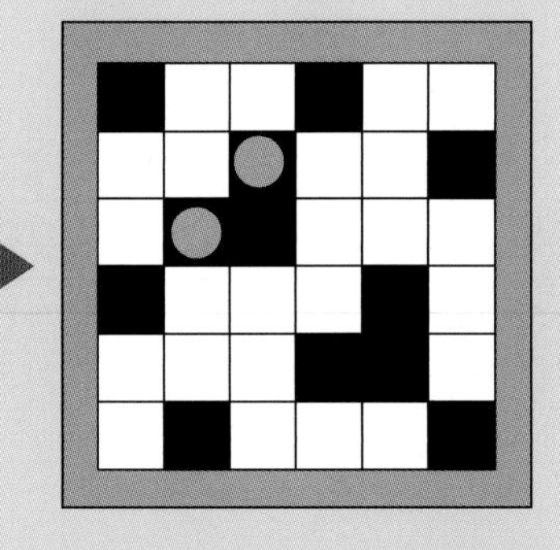

4

5
6
7
8
9

Some people get offended if you try to guess their age and get it wrong. (Some people might be offended if you guess correctly!) Practice with the people on these pages instead.

▼ TWO FATHERS

One father said to another: "If you multiply the ages of my four children, you get 39."

How old are the children?

ANSWER: PAGE 120

▼ FATHER AND SON

The man and his son have ages whose digits are the reverse of each other. The difference between their ages is 27.

How old are they?

ANSWER: PAGE 120

▲ LOST MARBLES

Jenny and Jemima each started out with the same number of marbles. After Jenny bought 35 more, and Jemima lost 15, they had 100 marbles between them.

How many marbles did they each have initially?

ANSWER: PAGE 120

▼ ARISTOTLE'S WHEEL PARADOX

A rolling wheel has many paradoxical properties. Points on its rim have different ground speeds as it rolls on its track. Where on the wheel are the maximum and minimum speeds reached?

The paradox was first mentioned in Mechanica, *a work attributed to Aristotle. The heart of the paradox was the following:*

As the large double wheel rolls from point 1 to point 2, the small wheel rolls from point 3 to point 4. If the double wheels roll along actual tracks, it is obvious they cannot roll smoothly along both tracks. If we assume that the wheel rolls smoothly without slipping from point 1 to point 2, at every instant a point on the big wheel can be put into one-to-one correspondence with a point on the small wheel, which is a paradox, because it seems to prove that the two circumferences have equal lengths.

How can this be explained?

ANSWER: PAGE 120

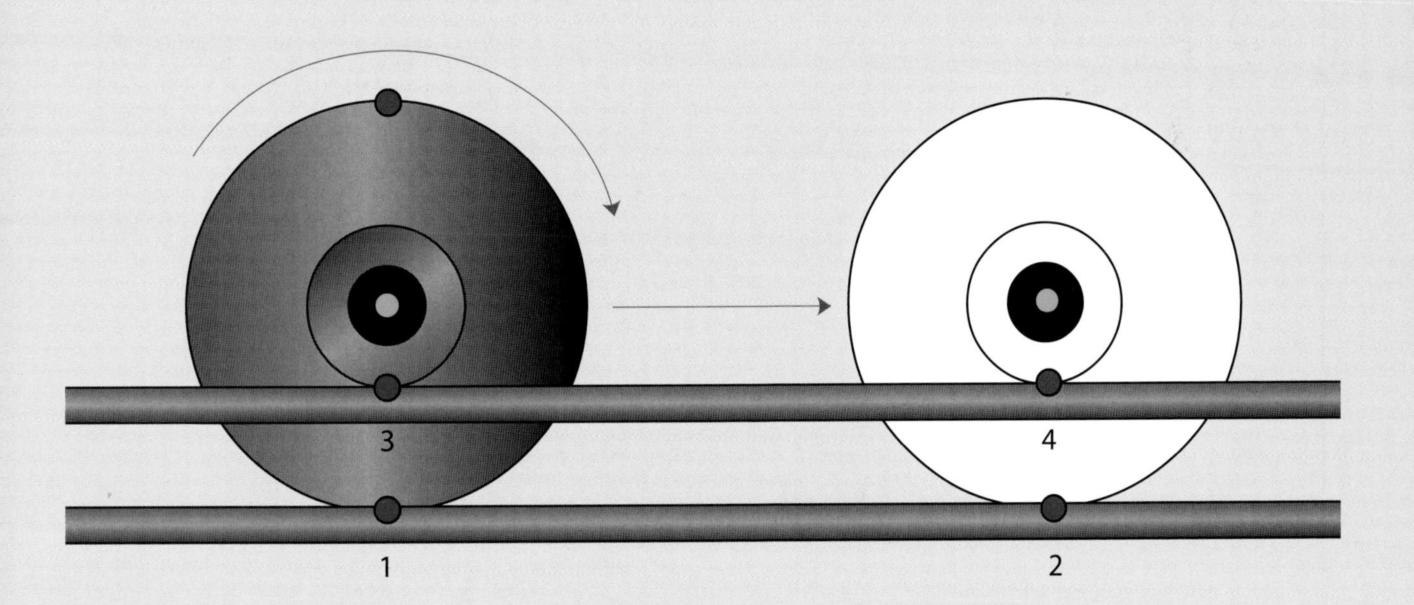

▶ LITTLE WOODEN MAN

Here is a tricky old puzzle.

A little wooden man stands on top of an old clock. Every time he hears the clock strike once he jumps twice. The clock strikes every hour, striking the number of the hours.

How many times does the man jump in 24 hours?

Answer: page 120

▼ INTEGRAL RECTANGLE

The big rectangle is dissected into smaller rectangles. Each of the smaller rectangles has either an integer height or an integer width. The green rectangles have integer widths and noninteger heights and the orange rectangles have integer heights and noninteger widths.

Does the big rectangle have integer width, an integer height, both, or neither?

ANSWER: PAGE 121

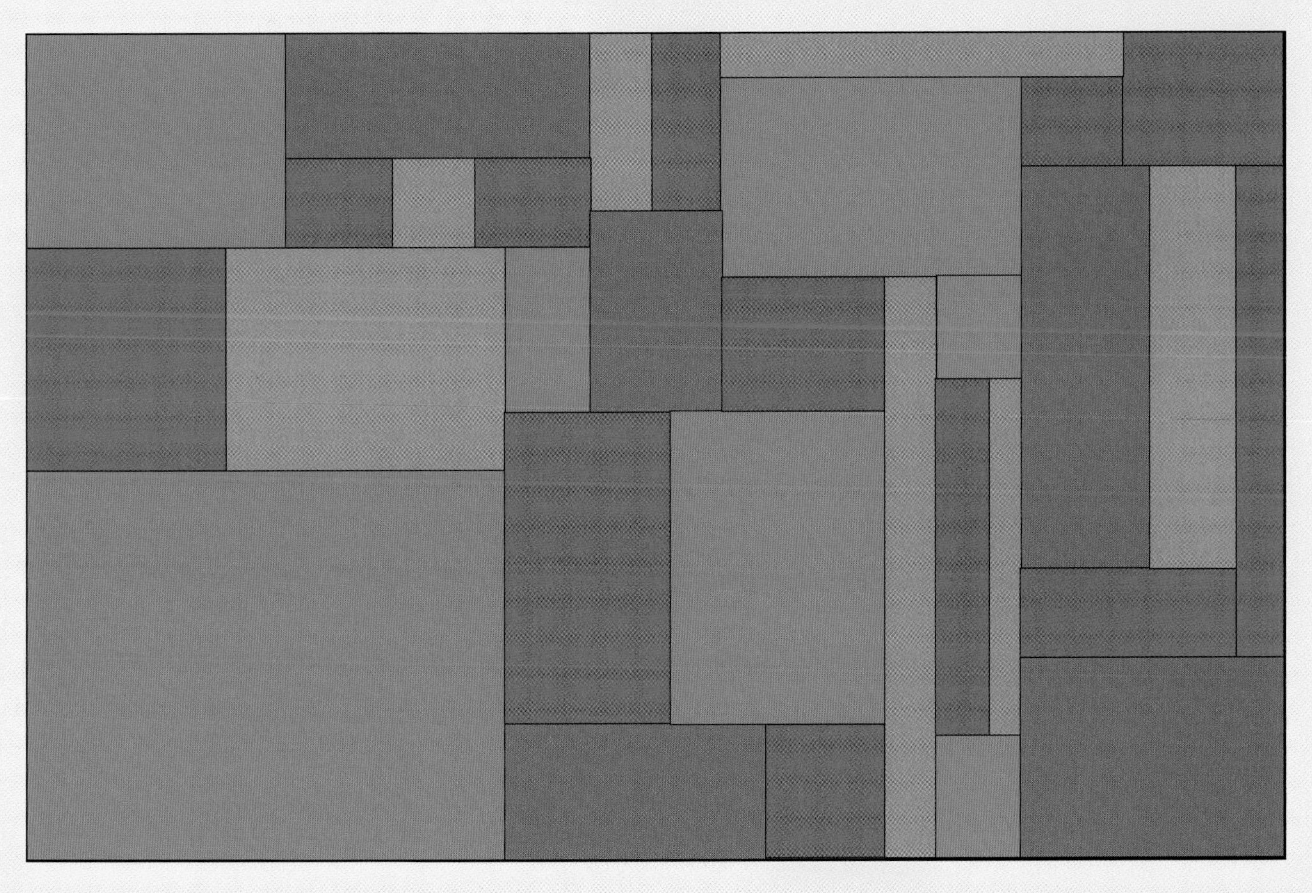

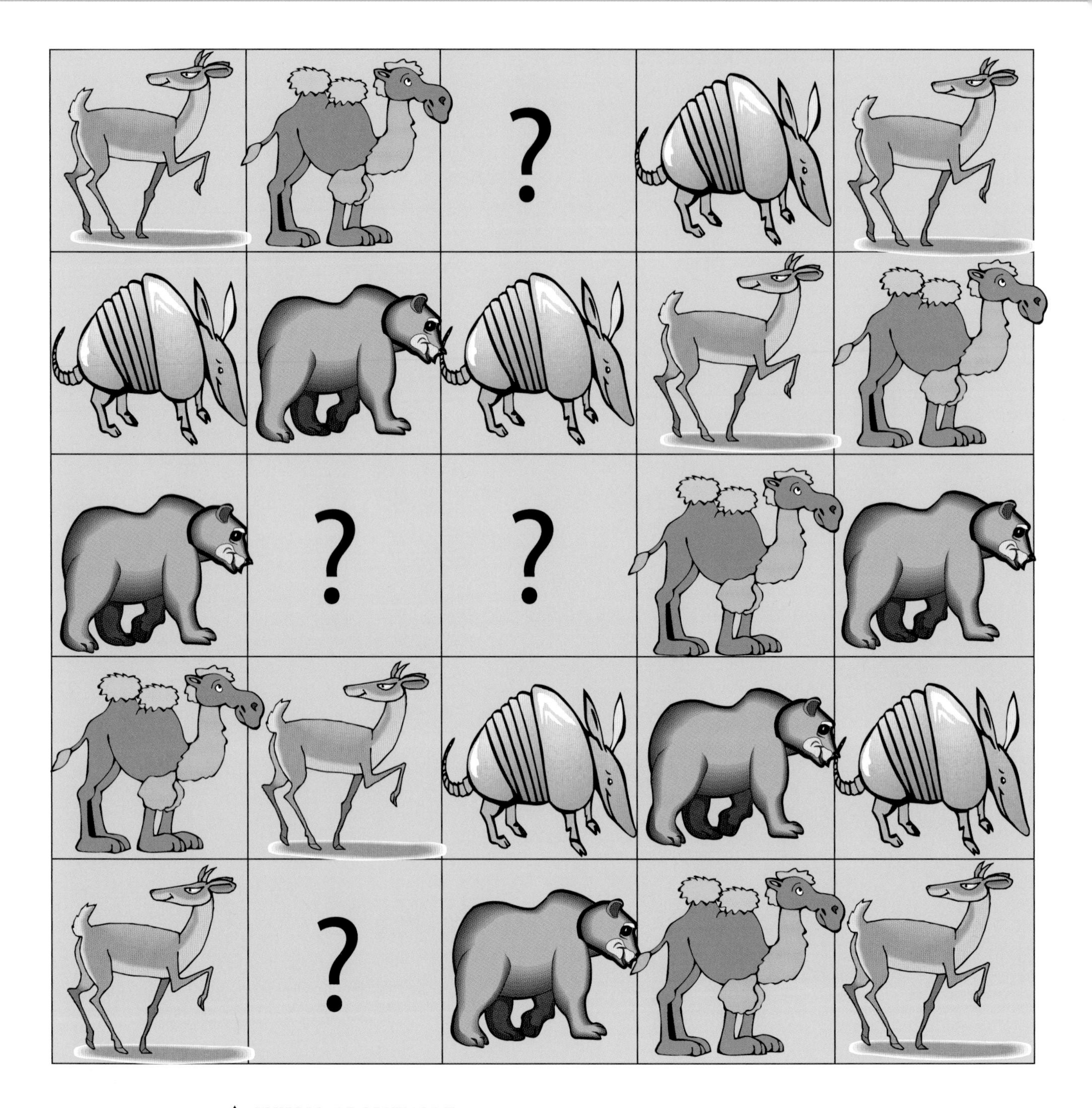

▲ ANIMAL PROMENADE

Which animals go into the empty squares?

Answer: page 121

▼ WHAT'S IN THE SQUARE?

All the shapes on the black squares around the periphery of the diagram appear in the corresponding rows or columns in a one-to-one correspondence.

There is one mistake in the pattern. How long will it take you to find and correct it?

ANSWER: PAGE 122

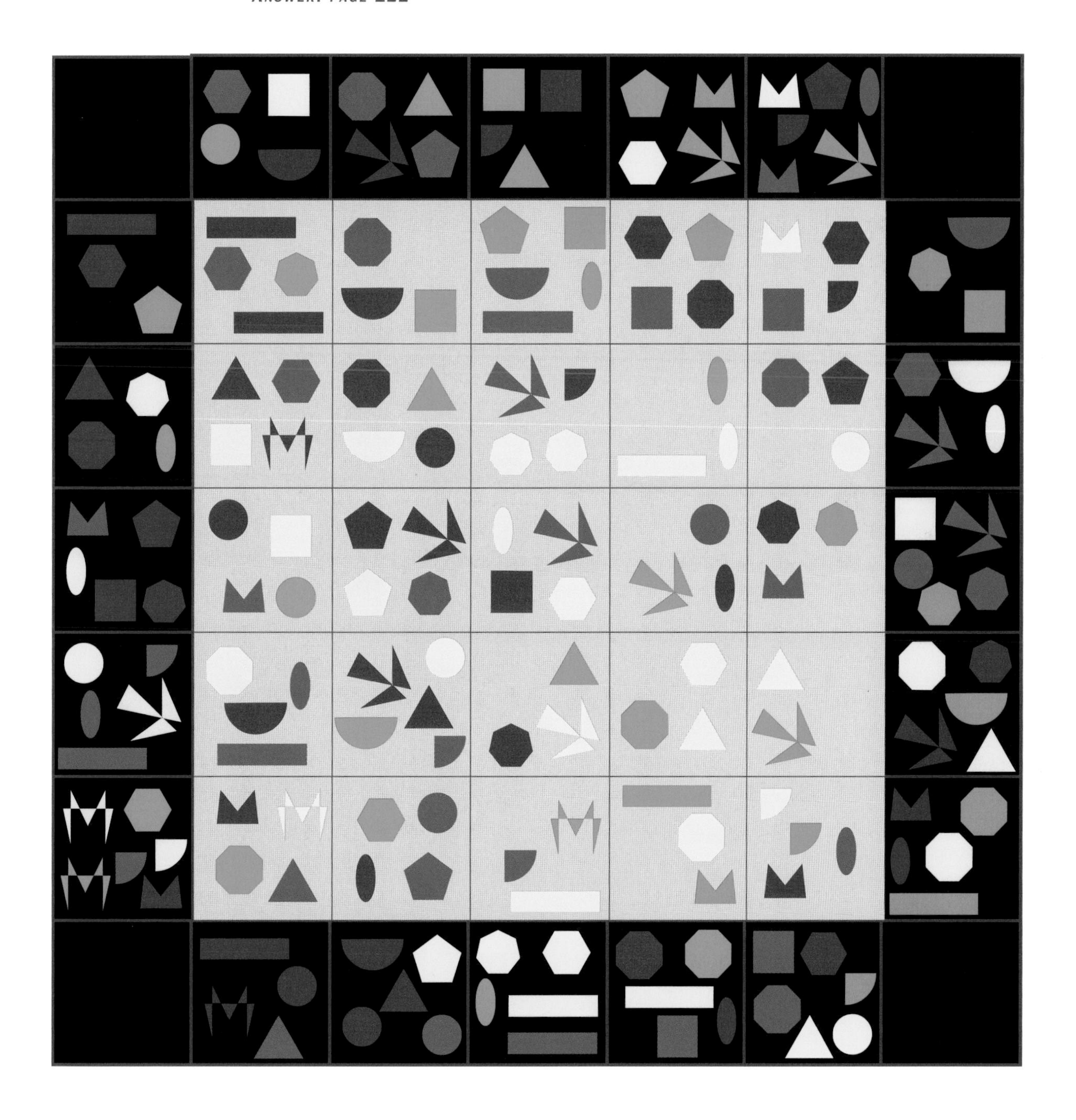

▲ SEVEN BIRDS

Seven birds live in a nest. They are very organized; each day three of the birds fly out in search of food.

After seven days every pair of birds will have been in exactly one of the seven daily search parties.

Number the birds from 1 to 7 and arrange a schedule for the seven search parties.

ANSWER: PAGE 122

Days	Group of three
1	
2	
3	
4	
5	
6	
7	

▲ WALKING DOGS

Nine girls, each with a pet dog, take their dogs for a walk every day. They walk in three groups of three, and over the course of four days, no pair of girls appears more than once among the groups of three.

Can you find a way for them to schedule this?

ANSWER: PAGE 122

Day 1			
Day 2			
Day 3			
Day 4			

▼ SCHEDULING SCHOOLCHILDREN

Fifteen schoolchildren go for a walk in groups of three for seven days in succession.

It is necessary to arrange them daily so that no two children walk together more than once during the seven days.

Can you fill in the table, scheduling the groups of children in threes for the seven days? Number the children from 1 to 15 for convenience.

There are seven basic solutions. Can you find one?

Answer: page 123

	TRIPLE S														
Day 1															
Day 2															
Day 3															
Day 4															
Day 5															
Day 6															
Day 7															

▲ CATS AND MICE CROSSING

Three cats and three mice want to get to the other side of a river. They have only one boat, which can hold only two animals. There can never be more cats than mice on either bank of the river, for obvious reasons.

Can they all cross safely?

What is the smallest number of trips in which the crossing can be completed?

ANSWER: PAGE 123

▲ PINWHEEL PATTERN

In every horizontal row and vertical column there are six pinwheels, each made of the same four colors.

Can you discover the logic of the pattern and color in the six blank pinwheels with the appropriate colors?

Answer: page 124

▲ WHERE IS TOM?

Tom always tells the truth. Dick sometimes tells the truth and sometimes lies. Harry always lies.

Which is which?

ANSWER: PAGE 124

1) Exactly one of these twelve statements is false.

2) Exactly two of these twelve statements are false.

3) Exactly three of these twelve statements are false.

4) Exactly four of these twelve statements are false.

5) Exactly five of these twelve statements are false.

6) Exactly six of these twelve statements are false.

7) Exactly seven of these twelve statements are false.

8) Exactly eight of these twelve statements are false.

9) Exactly nine of these twelve statements are false.

10) Exactly ten of these twelve statements are false.

11) Exactly eleven of these twelve statements are false.

12) Exactly twelve of these twelve statements are false.

▲ TRUE STATEMENT

Which of the above twelve statements is true?

ANSWER: PAGE 124

▲ **TRUTH CITY ROAD**

The inhabitants of Truth City always speak the truth, while those of the city of Liars, of course, always lie.

On your way to visit Truth City, you arrive at the crossroads leading to the two cities. As you can see, the sign is confusing, so you are forced to ask a man standing at the crossroads for the right direction.

Unfortunately, you don't know whether the man is a citizen of the city of Liars or is from Truth City. You are allowed to ask the man only one question.

What question can you ask him to be sure to learn how to get to Truth City?

ANSWER: PAGE 124

▲ TRUTH, LIES, AND IN BETWEEN

The citizens of the cosmopolitan city Nooneknowstruth are of three types: those who always speak the truth, those who always lie, and those who alternately lie and tell the truth.

Again, you meet one of the residents. This time you are allowed to ask two questions. His answers must be sufficient for you to determine to which of the three groups the man belongs.

What are the two questions you will ask him?

ANSWER: PAGE 124

▲ TRUTH AND MARRIAGE

The king has two daughters, Amelia and Leila. One of them is married, and the other not. Amelia always tells the truth, while Leila always lies. A young man is allowed to ask just one single question of one of the daughters to find out which daughter is the married one. His reward would be, of course, marrying the unmarried daughter.

The catch is, his question may not contain more than three words. Also, he does not know which daughter is which.

What can he ask to win a bride?

Answer: page 125

▶ SWIMMING POOL

Could you drink enough water in your lifetime to empty a swimming pool that is 10 meters long, 5 meters wide, and 1 meter deep?

Answer: page 125

▲ FIGARO

In a small town, the only barber is named Figaro. Some residents have beards and do not shave, and still others shave themselves. Of those who do not have beards, however, Figaro shaves all who do not shave themselves. He shaves none of those who do shave themselves, for every man either shaves himself, or else Figaro shaves him, and no man adopts both methods. The question is, does Figaro have a beard?

ANSWER: PAGE 125

Fortunately, our lives do not depend on getting the correct answer to this puzzle. But do you know whether these prisoners of war will live or die?

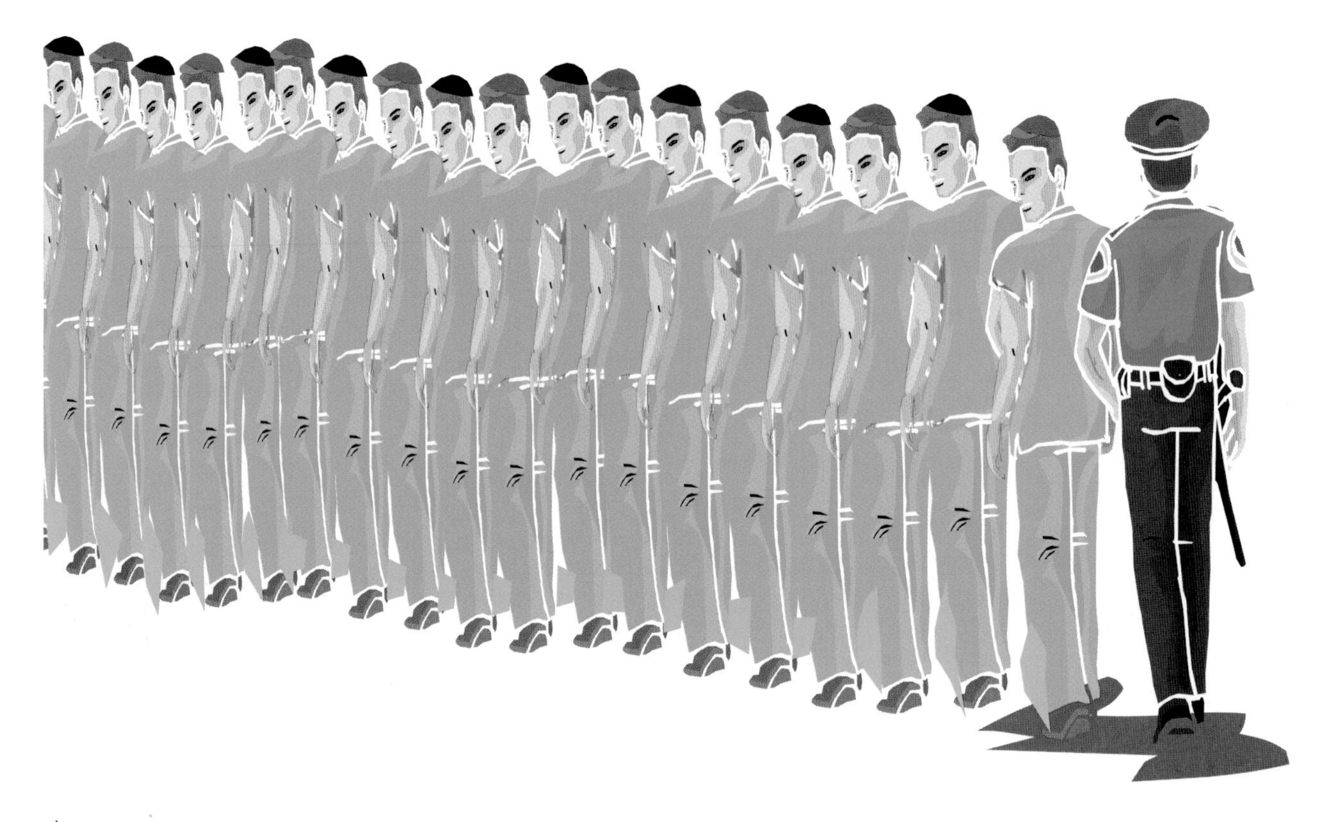

▲ HATS AND PRISONERS

During the Second World War there were 100 prisoners in a POW camp. The guards wanted to take a vacation, and one of their suggestions was to get rid of the prisoners by shooting them all. The camp commander, of a more sporting nature, agreed to the idea, but decided to tell the POWs that they would be shot unless they could answer one question.

So all the prisoners were gathered and he said:

"You are all dirty dogs, and I intend to have all of you shot. But being a fair man, I am going to give you one last chance to live. You will be taken to the dining hall, where I shall arrange for a large crate containing the same amount of red and black hats to be delivered. You will leave one by one. A hat, randomly picked from the crate, will be placed on your head. You will not be able to see the color of your hat, although you will be able to see the others. You will form a single line and you will be shot on the spot if you talk or signal to one another".

"I shall walk down the line asking each of you for the color of the hat you are wearing. If you answer correctly, you will be set free. If not, you will be shot".

The prisoners were ushered into a big hall, where they could discuss the situation and try to come up with some kind of a strategy to cope with it.

Some time later, each prisoner got his hat, and the commander, expecting to shoot at least 50% of the prisoners, started asking them for the color of their hats.

Can you imagine his anger and frustration when he found that he had to set free...well, how many prisoners did *he have to set free?*

ANSWER: PAGE 125

▲ MY CLASS

In a group of 20 boys, 14 have blue eyes, 12 have black hair, 11 are overweight, and 10 are tall.

How many boys share all four of these features?

ANSWER: PAGE 126

▼ COCKTAIL AT LARGE

Which is greater—the circumference of the cocktail glass, or its height?

ANSWER: PAGE 126

▲ HOTEL DOORS

Ten hotel doors are numbered from 1 to 10, all closed.

A cleaning lady walks by and opens every second door starting with door 2.

Then a repairman walks by and changes the state of every third door (opening it if it is closed, and closing it if it is open).

Then another person walks by and changes the state of every fourth door, etc. This goes on until no more doors can be altered.

Which doors will be closed at the end?

ANSWER: PAGE 127

▶ HATS AND COLORS 1

Four clowns are wearing two red and two green hats. They know that there are two hats of each color, but none of them knows the color of his own hat, and they are not allowed to turn and look behind them.

Which of them will be the first to shout out the color of his hat? Note: The first clown cannot be seen by the others, since he is obstructed by the circus poster.

ANSWER: PAGE 127

▶ HATS AND COLORS 2

Three red hats and two blue hats are worn by the five clowns. They know how many hats there are of each color, but none of them knows the color of his own hat. As in the previous puzzle, they are not allowed to turn and look back.

Additionally, clown E can be seen only by clown D. We see the clowns from clown A's perspective, so we, like clown A, do not know what color hat is worn by clown A or clown E.

Which of the clowns will be the first able to say aloud the color of clown A's hat?

ANSWER: PAGE **128**

▼ ALHAMBRA PATTERN

The former palace of the Moorish kings of Granada is a place of mathematical beauty. The intricate pattern shown below is an example of the remarkable wealth of its many complex geometrical designs.

Can you tell whether the pattern is made of a single loop or not? If it's composed of separate parts, how many are there?

Answer: page 128

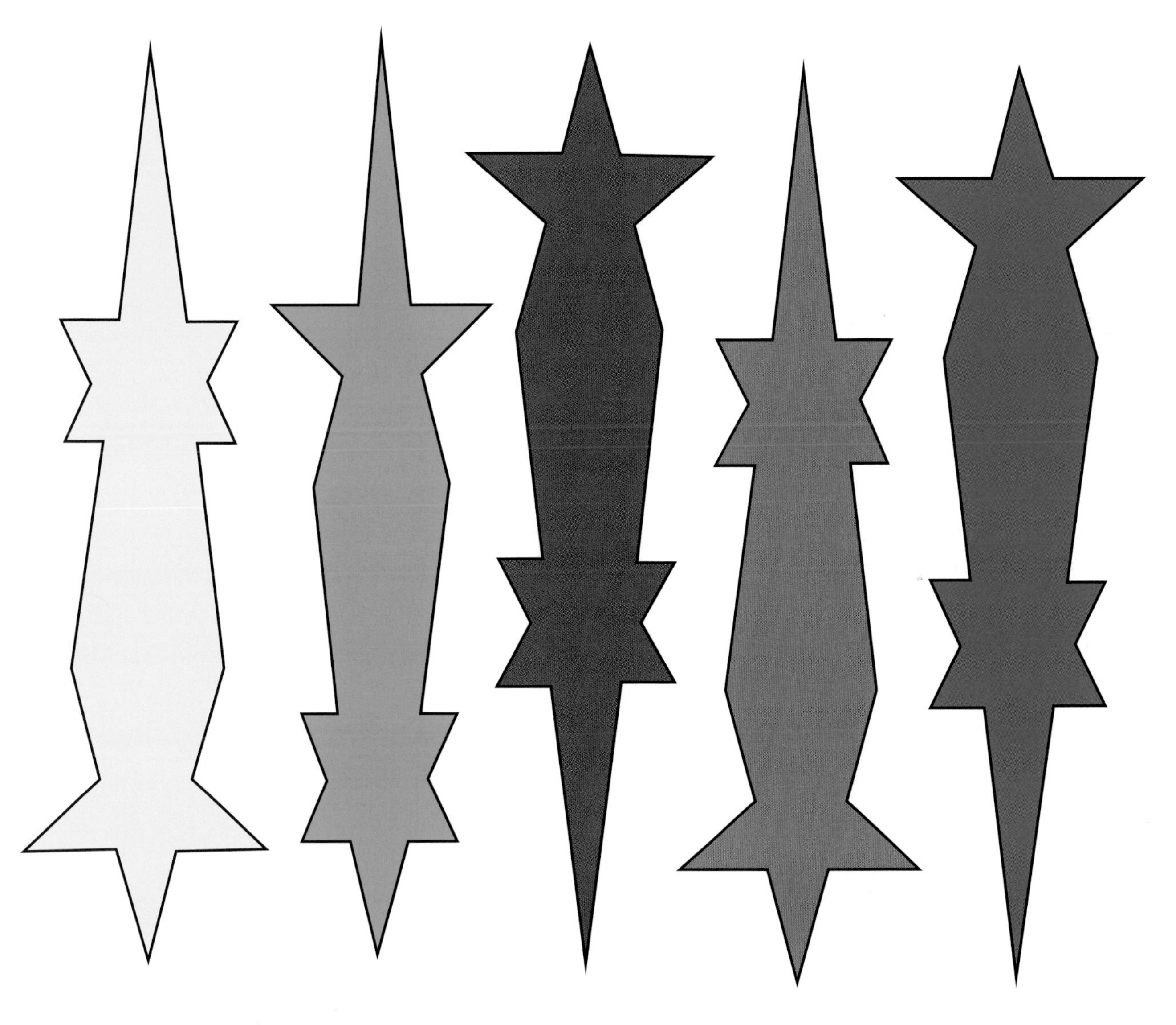

▲ ODD ONE OUT

Which of these shapes is different from the others?

ANSWER: PAGE 128

UP OR DOWN? (PAGE 6)

1) up
2) down
3) up
4) down

SNEEZE (PAGE 7)

You will travel about 48 feet with closed eyes before you can brake your car, so you will just barely avoid an accident.

There are 1760 yards in a mile, so at 65 miles per hour you would travel 65 × 1760 × 3 feet. Divide this by the number of seconds in an hour (60 × 60) then by 2 again (for half a second). This gives us 47.66ft, which is enough space to prevent a collision.

INCLINED PLANE (PAGE 8)

A ball that rolls a distance *d* down an incline in 1 second rolls 4 times as far in 2 seconds, 9 times as far in 3 seconds, and 16 times as far in 4 seconds. You can easily test this by rolling a ball on a ruler, provided the angle of incline is sufficiently small, so that the ball will remain rolling for as long as 4 seconds.

RACE (PAGE 9)

When runner A reached the finish line, runner B was at the 90-yard mark; he covers only 90% of the distance covered by runner A in the same amount of time. Similarly, runner C runs at 90% of the running speed of runner B, so he would only be at the 81-yard mark when runner B is at the 90-yard mark.
Thus A beat C by 19 yards.

REFLEX RULER REACTION (PAGE 10)

With the reflex ruler you can test the reaction time of yourself and of your friends. The ruler is based on the fact that the distance traveled from the time the ruler is released is proportional to the square of the time that it took the ruler to be caught.

Hold the ruler above and between the thumb and forefinger of your friend to be tested, at the circle spot at the bottom of the ruler.

Release the ruler without warning. The mark where your friend catches the ruler will show the score.

MALTESE CROSS MECHANISM (PAGE 11)

The machine shown is a film projector, which displays one frame of film at a time. Each frame is lit for a fraction of a second before the projector's lamp goes off, the next frame is put into position, and the new frame is shown. Due to persistence of vision, our eyes don't notice the gap between the frames. When you go to a typical film, you are watching a blank screen for over an hour!

The frames must be shown individually to avoid blurring, which is why the Maltese cross system is ideal.

Here's how it works: The driving wheel continuously rotates. Each time the pin (A) enters one of the four slots (B) on the Maltese cross (C), it gives the cross a quarter turn, rotating the wheel (D) that holds the film, one frame of film at a time. The notch in the yellow wheel attached to the driving wheel allows the cross to rotate; the rest of the time the cross is held in place by the yellow wheel.

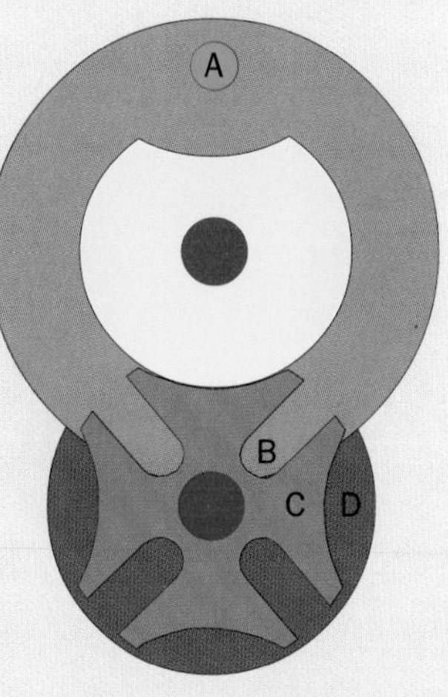

RATCHET-WHEEL MECHANISM (PAGE 12)

With every clockwise revolution of the green driving wheel, the arm will move the ratchet wheel one tooth counterclockwise.

SLIDING-FRAME MECHANISM (PAGE 13)

Continuous revolving motion of the red wheel results in an alternating to-and-fro sliding motion of the blue frame.

SWINGING PENDULUMS (PAGE 15)

The pendulum will swing back and forth in the same amount of time. It's an astonishing fact that the time of a pendulum swing varies not with the size of the swing but only with the length of the pendulum itself, which is quite counterintuitive. Whether it makes a long swing or a short one, the period will be the same. With this simple observation Galileo invented the pendulum clock.

The strange motion of a pendulum obeys certain laws:

1) The period of oscillation does not depend on the weight of the bobs.
2) The period does not depend on the distance traveled.
3) The period of oscillation is proportional to the square root of the length of the pendulum.

A pendulum's period, or the time (T) it takes to go through one cycle, can be expressed by the simple formula: $T = \sqrt{\frac{L}{g}}$ where L is the length and g is the rate of acceleration due to gravity, which is 9.80 m/sec^2.

Since g is the only variable besides the length, a pendulum is a simple way to measure the gravity of particular planets. A 1-yard-long pendulum will complete a swing in about 1 second on earth and 2.5 seconds on the moon.

The thing that's wrong with our problem is the clock pendulum taking 5 seconds for a swing. A pendulum that takes that long to swing would be too long to fit in any living room!

We are assuming that there is no friction. If the pendulum was quite light in a normal air-filled room, over time the pendulum would lose its energy.

Generally, the pendulum theory works best for smaller angles. If it swung through 140 degrees, other physical factors would start to make an impact.

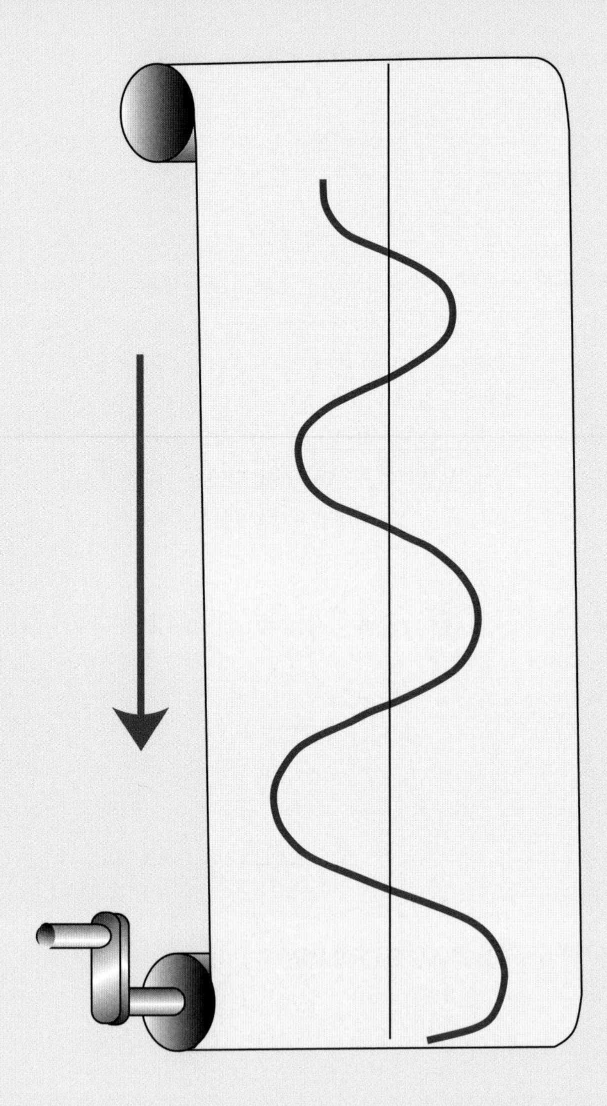

◀ SIMPLE HARMONIC MOTION (PAGE 16)

The outcome of the motion of the pendulum with the attached pen will produce a sinusoidal curve like the one shown. The motion is called damped harmonic motion because of the friction which will ultimately stop the swinging pendulum, producing a straight line.

The ideal case (when there is no friction) is called simple harmonic motion. Simple harmonic motion is one of the commonest type of motions in nature, from ripples in a pond to radio waves, etc.

COUPLED PENDULUMS (PAGE 17)

The motion of the pendulums is coupled by means of the connecting string. When one pendulum is made to swing, it moves the connecting string, which then moves the other pendulum. Its energy will gradually be transferred to the other pendulum, and back again.

Because of this resonance transfer, coupled pendulums of this type are often called resonance pendulums.

RESONANCE (PAGE 18)

After a while, all the pendulums will start swinging, but only the matching pair will attain a noteworthy amplitude, and they will exchange their energy in an oscillatory fashion.

Every pendulum has a resonant or natural frequency. Every time a pendulum swings it pulls the connecting rod and gives the other pendulums a small pull. The pull of the pendulum on another having the same length will be at the natural frequency of the second pendulum, which will begin to swing too.

Eventually, the first pendulum will be brought to rest when it has transferred all its energy to the second pendulum, and then everything reverses and starts all over again.

ONE-TON PENDULUM (PAGE 19)

By many repeated small pulls on the string, the massive pendulum will slowly start swinging in larger and larger arcs—as long as the rhythm of repetition of the small pulls is resonant.

If you pull too strongly the thread will break. By gently pulling the string when it is attached to the pendulum, the pendulum will move a bit. Let it swing, then throw the magnet again and pull it back just at the moment when the pendulum ends its swing toward you. If you always pull rhythmically at the right moment, the oscillations of the pendulum will increase—little effort, great effect.

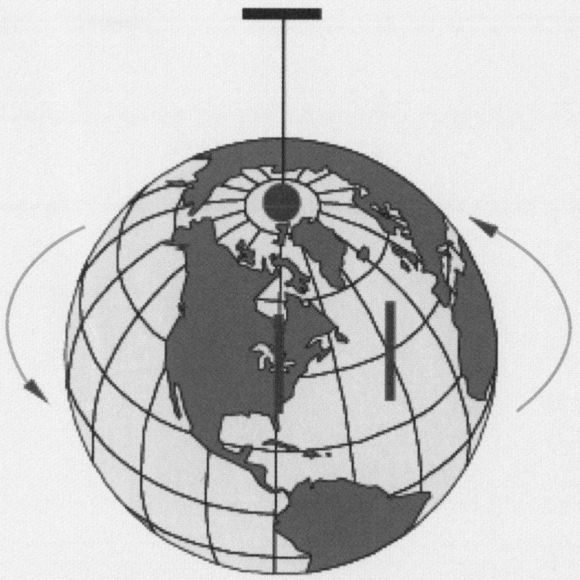

◀ FOUCAULT'S PENDULUM (PAGE 20)

The swinging pendulum continues to swing through the same plane once it is set in motion. Its changing paths in the sand below can be explained only by the fact that the Earth itself turns beneath the pendulum.

The apparent rotation of a pendulum varies with the latitude at which it is installed. Its rate at points between the poles and the equator is equal to 15 degrees per hour multiplied by the sine of the latitude.

GROWTH AND SIZE (PAGE 22)

You would weigh eight times as much.

When we double the linear measurements of a two-dimensional object the area is multiplied by four (2×2).

Similarly, for a three-dimensional object, since each of the three dimensions is doubled, the volume, and thus the weight, is multiplied by eight ($2 \times 2 \times 2$).

STAIRCASE PARADOX (PAGE 23)

In the 10th generation there will be $2^{10} = 1024$ stairs.

As for the length of the staircase, all the staircase paths in all generations are of the same length—twice the side of the unit square = 2.

On the other hand, as the progression continues, the steps will, to the naked eye, become indistinguishable from the diagonal of the square, whose length is, according to the Pythagorean theorem: $\sqrt{1^2 + 1^2} = \sqrt{2}$

It looks like we have just proved a paradoxical relationship ($2 = \sqrt{2}$), haven't we?

While it appears that the small staircases approach the length of the diagonal, in fact they don't. The steps grow smaller and more numerous, but they always add up to 2. The steps may become too small to see without magnification, but they're there.

INFINITY AND LIMITS (PAGE 24)

The pictures would approach twice the height of the original picture, but would never reach that size, no matter how long you go on with the sequence:

$1 + \frac{1}{2} + \frac{1}{4} + \frac{1}{8}\ldots$

GNOMONIC EXPANSION (PAGE 25)

The red gnomon's area is a^2, or $\frac{1}{4}$ of the bottom square.

The circular arc shows that the length of a side of the inner square at the top is equal to the length of a side of the entire lower square. Therefore, the red gnomon's area is equal to the area of the small square on the right.

ARITHMETICAL PROGRESSIONS (PAGE 26)

For some simple arithmetical progressions, the common difference can be discovered by a surface level analysis, but for some cases a multilevel analysis is required before the common difference is found. This is demonstrated in the solutions to the two problems:

Puzzle 1) 20 28 40 56 76 level 0
8 12 16 20 level 1
4 4 4 level 2

By adding 4 + 16 + 56 = 76 the next number is found.

Puzzle 2) 8 26 56 100 160 238 336 level 0
18 30 44 60 78 98 level 1
12 14 16 18 20 level 2
2 2 2 2 level 3

By adding 2 + 18 + 78 + 238 = 336 the next number is found.

It should be noted that a multilevel analysis will not always provide a common difference, and other ways must be found in such cases to find it. Some sequences contain common factors rather than a common difference, and successive terms are found by multiplying by a fixed number.

A sequence of numbers that have a common factor is called a geometrical progression. Common factors are found by dividing a successive term by its predecessor:

2 6 18 54
6 ÷ 2 = 3 18 ÷ 6 = 3 54 ÷ 18 = 3

Therefore the next number in the sequence is 54 x 3 = 162.

▼ ROW OF SKYSCRAPERS (PAGE 27)

Two solutions are shown here. Among the 9! (nine factorial) or 362,880 different permutations of the nine skyscrapers, there are 84 arrangements which comply with the requirements.

WATER LILIES (PAGE 28)

59 days; on the second day of the example given the lake has two lilies.

SNOWFLAKE AND ANTI-SNOWFLAKE CURVES (PAGE 29)

It is easy to prove that the area of the curve is finite. The fact that the curve appears to remain within the page of our book is a good indication. At no stage of the development will the curve extend beyond the circle circumscribed about the initial triangle. The limit of this infinite construction closes an area $\frac{8}{5}$ that of the original triangle.

Now about the length of this curve. Suppose that the side of the original triangle is 1 unit long, then the perimeter is 3 units. In the construction of the second polygon each segment is replaced by two line segments that altogether, are equal to $\frac{4}{3}$ its length. Thus, at each stage, the total length is increased by a factor of $\frac{4}{3}$. Clearly this is not bounded. A curve of infinite length is the result.

An important principle shown by the snowflake and similar pathological curves is that complex shapes can result from repeated applications of very simple rules. These shapes are called fractals. The snowflake curve was discovered in 1904 by Helge von Koch.

Are there three-dimensional analogs of the snowflake and similar curves? For example, if tetrahedrons are constructed on the faces of tetrahedrons, will the limiting solid have an infinite surface area? Will it enclose a finite volume?

Can you imagine a shape which has an infinite length yet encloses only a finite area? It sounds impossible, but as we have shown, such figures exist.

SQUARES IN SQUARES (PAGE 30)

If the process is repeated infinitely the total area of the gold squares increases until it becomes the area of the initial square, a surprising and counterintuitive outcome—but such outcomes are not uncommon when dealing with infinity.

First stage: 1 gold square of area $\frac{1}{9}$ = total area 0.111
Second stage: 8 gold squares of area $(\frac{1}{9})^2 + 0.111$ = total area 0.209
Third stage: 8^2 gold squares of area $(\frac{1}{9})^3 + 0.209$ = total area 0.297
Fourth stage: 8^3 gold squares of area $(\frac{1}{9})^4 + 0.297$ = total area 0.375.

The pattern becomes clear. The total sum of gold areas is the infinite sum:

$$\frac{1}{9} + 8 \times (\frac{1}{9})^2 + 8^2 \times (\frac{1}{9})^3 + 8^3 \times (\frac{1}{9})^4 + \ldots$$

If we follow this series to the 25th stage, the total area of the gold squares is 0.947, and it becomes clear that this sum is steadily approaching 1, which is the initial area of the blue square.

▶ SIERPINSKI TRIANGLE (PAGE 31)

The fourth generation of the Sierpinski triangle. The proportions of the black triangles to whole triangle:

First generation: 25%

Second generation: ~ 44%

Third generation: ~ 58%

Fourth generation: ~ 68%

If you go on dividing the white triangles according to the rules, the white area will constantly decrease, approaching zero as the limit.

Fourth generation
$175 \div 256 = 0.68$

▼ FIBONACCI-SQUARE PROGRESSION (PAGE 32)

The fifth generation of the Fibonacci squares progression is shown below. As the series progresses, the proportions of the black areas in relation to the whole square increases; in this generation the ratio is about ⅔.

▼ SPIROLATERALS 1 (PAGE 34)

There is no closure in the series for orders that are divisible by four.

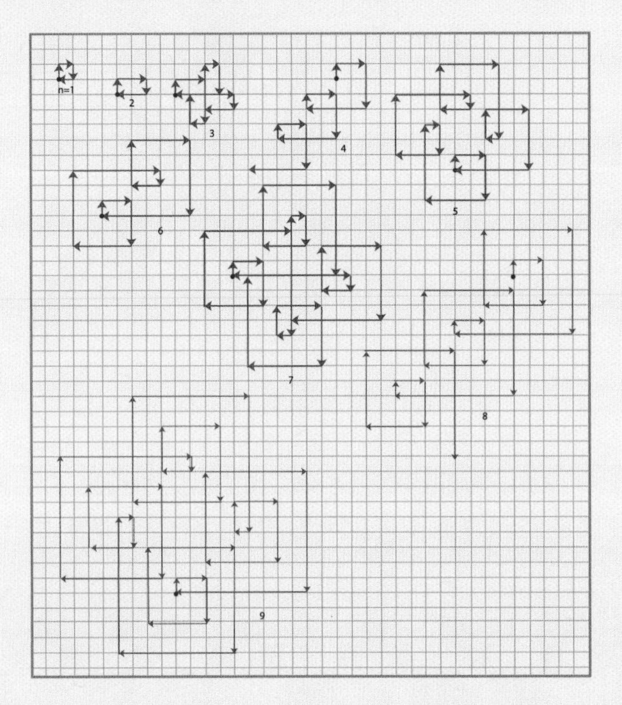

▼ SPIROLATERALS 2 (PAGE 35)

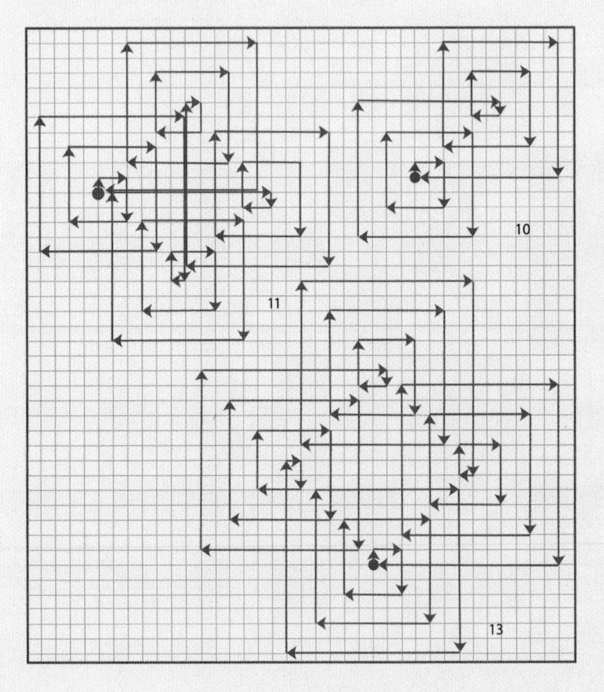

▼ SPIROLATERALS 3 (PAGE 36)

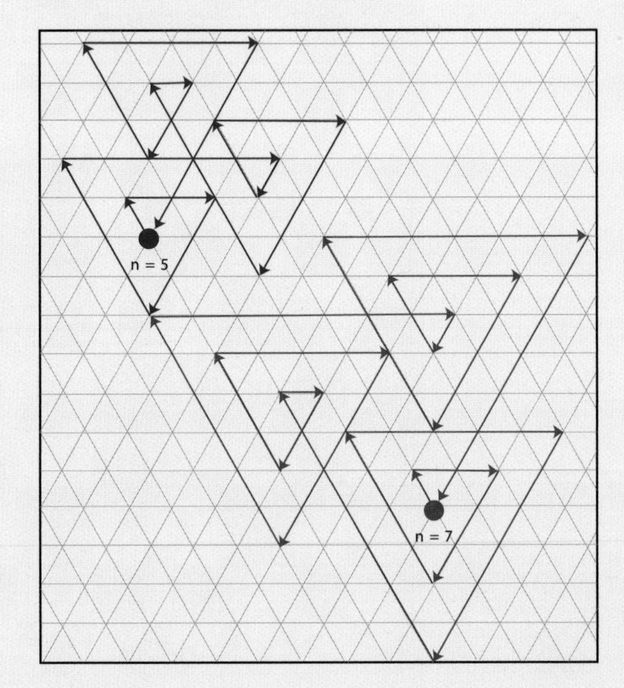

▼ SPIROLATERALS 4 (PAGE 37)

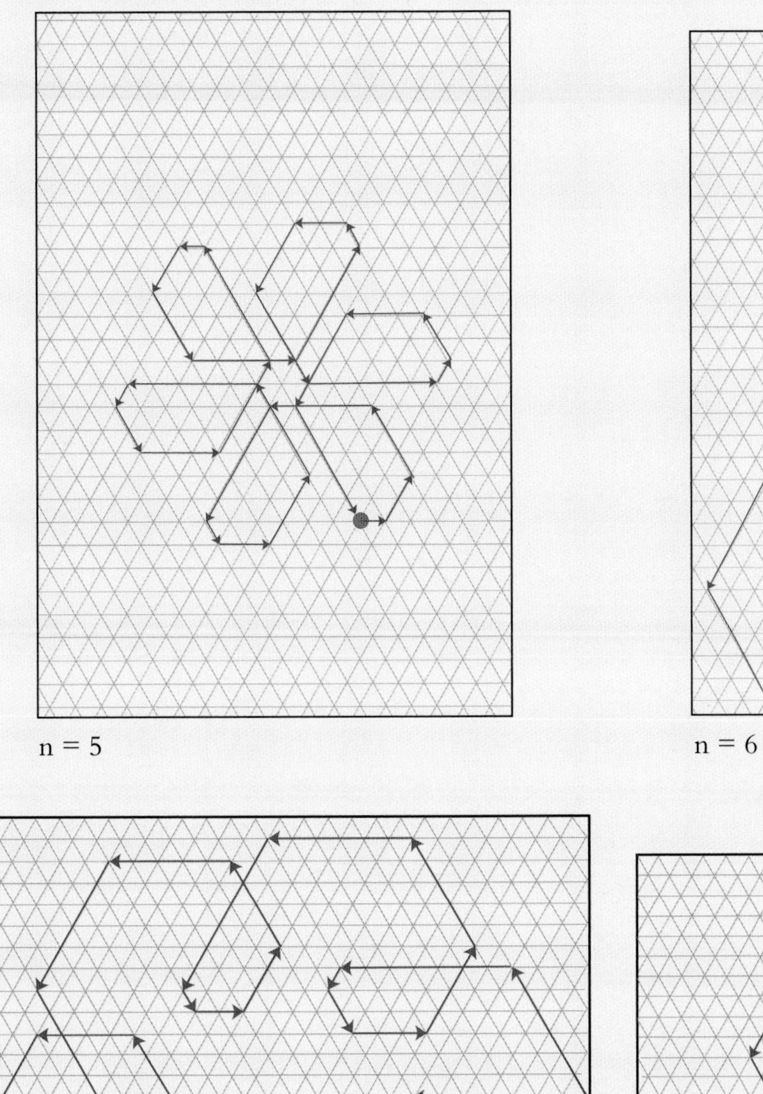

n = 5

n = 6

n = 7

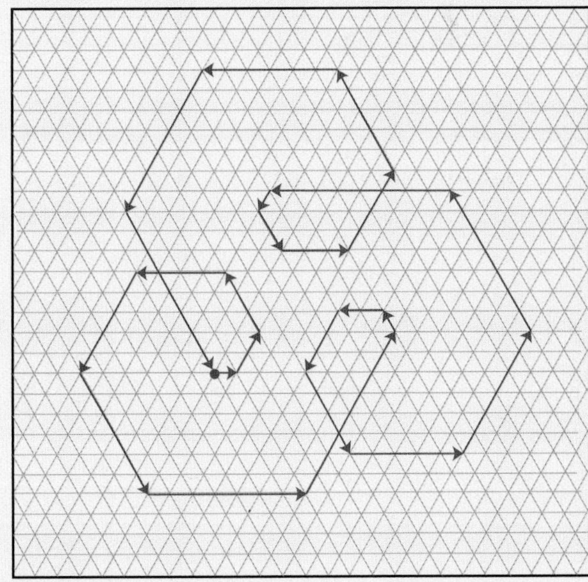

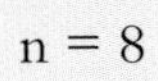

n = 8

▼ LONGEST PATH (PAGES 38–39)

Puzzle 1) A solution in five moves. No solution is possible with more than five moves.

Puzzle 2) An eleven-move solution seems to be the longest possible. Can you do any better if the line is allowed to cross itself?

Puzzle 1

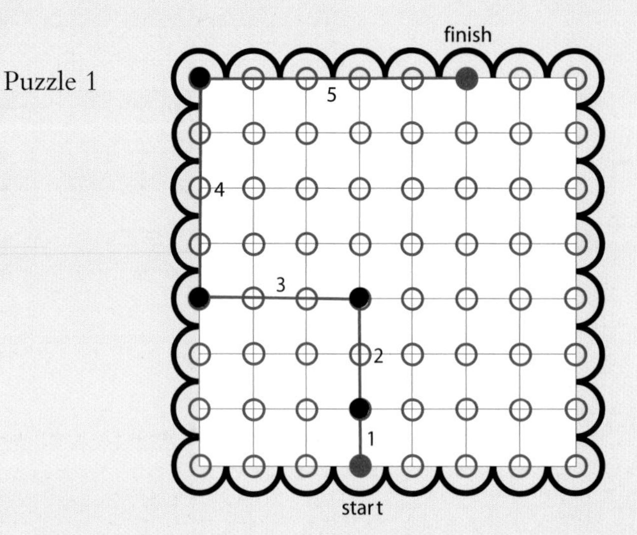

Puzzle 2

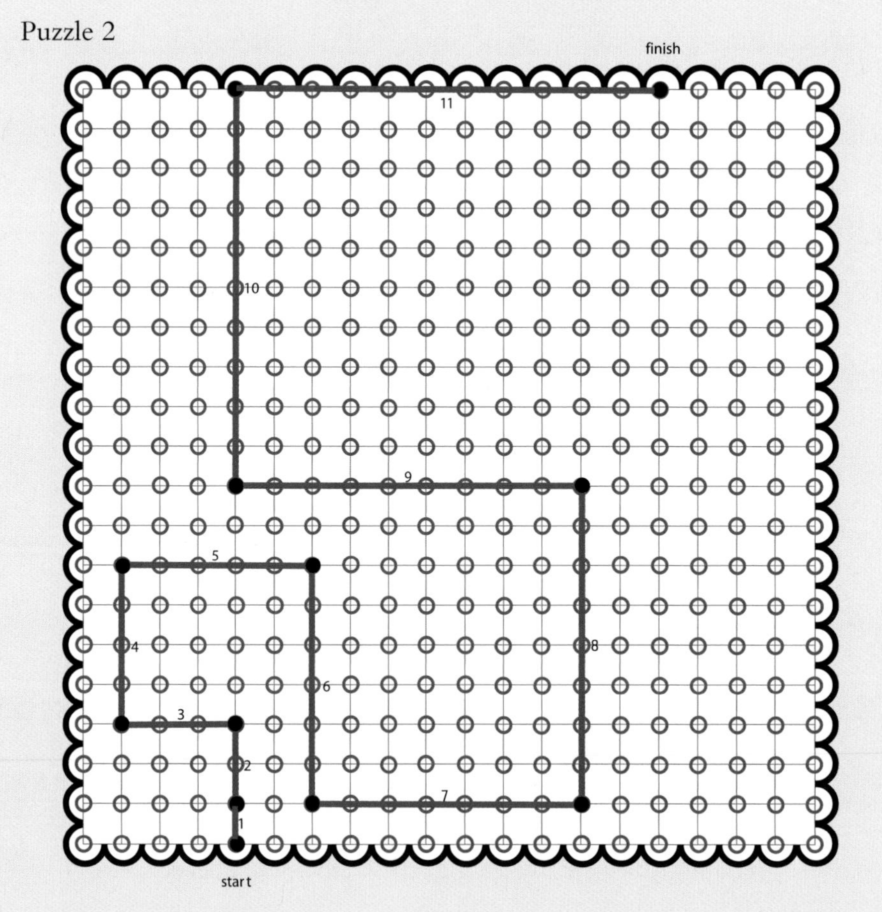

▼ CELLULAR PATHS 1 (PAGES 42–43)

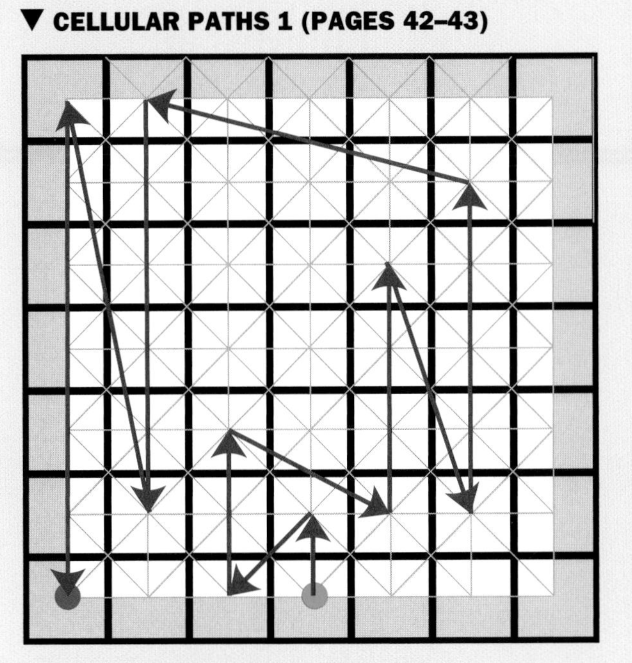

Puzzle 1) 7 x 7 square
11 moves

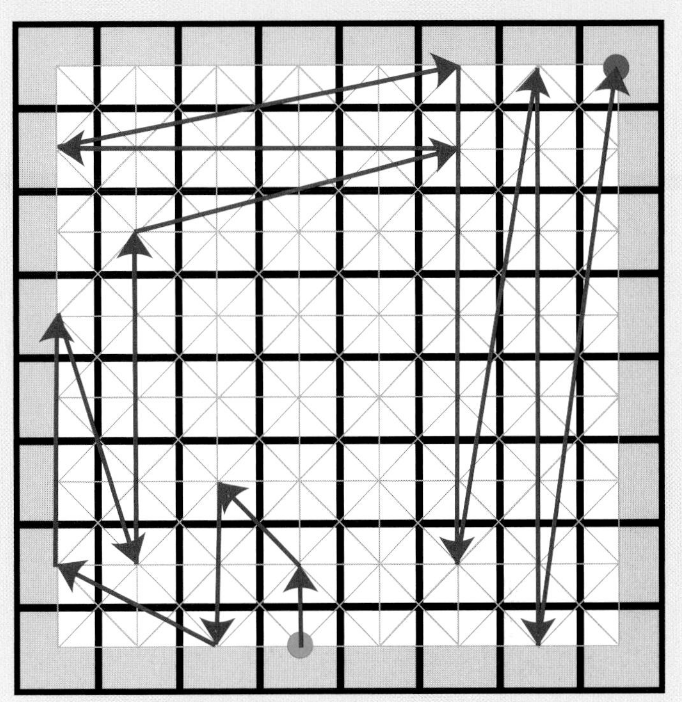

Puzzle 2) 8 x 8 square
14 moves

▶ CELLULAR PATHS 2 (PAGE 44)

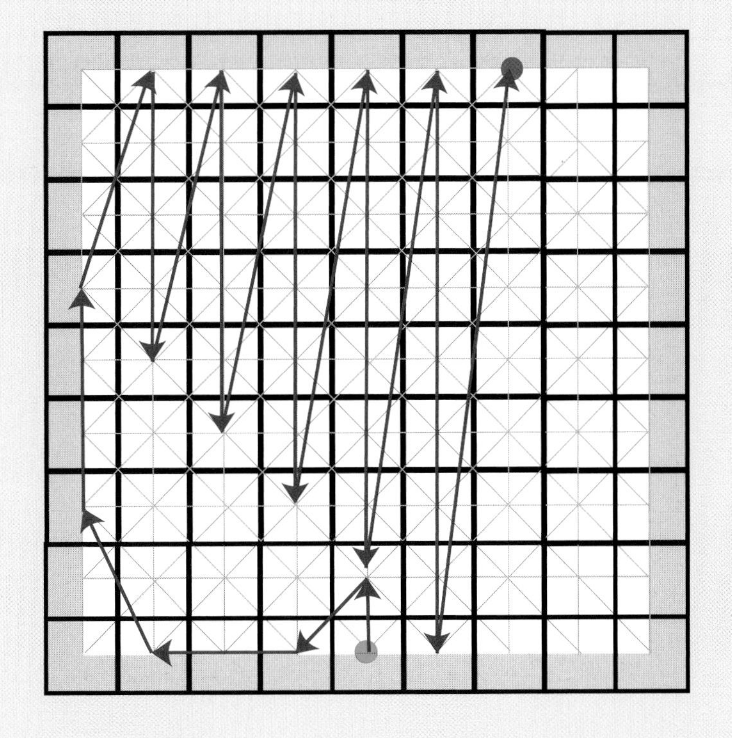

9 x 9 square
16 moves

▼ CELLULAR TRANSFORMATION (PAGE 45)

The pattern repeats after the third generation.

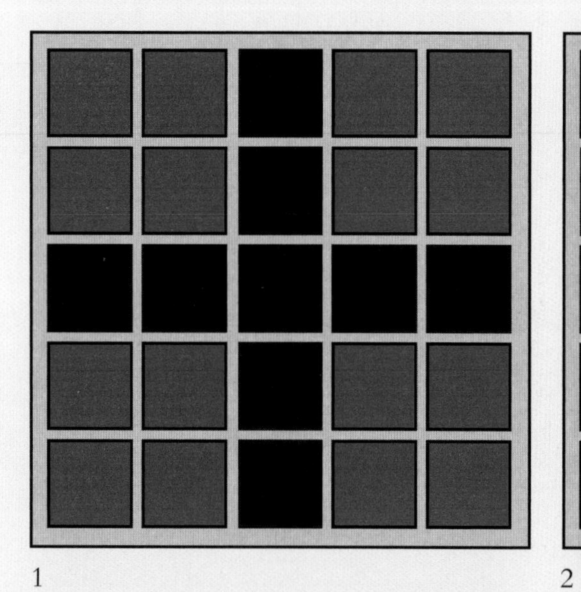

1

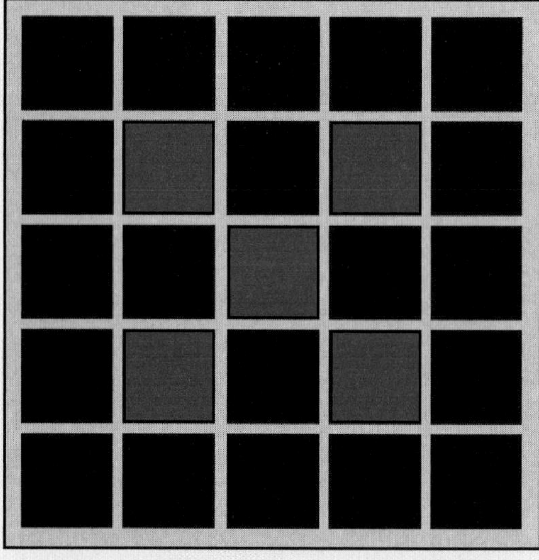

2

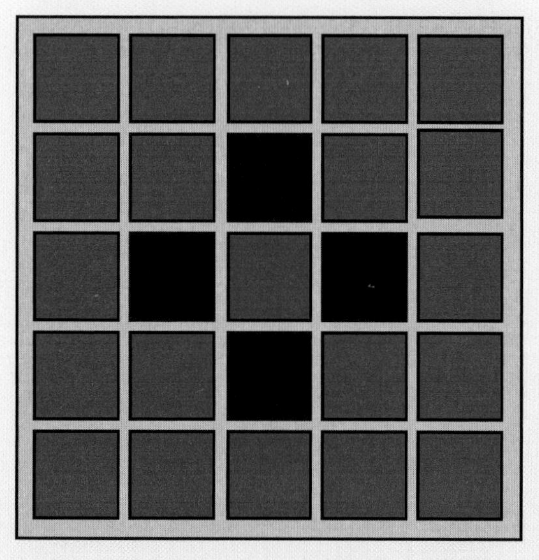

3

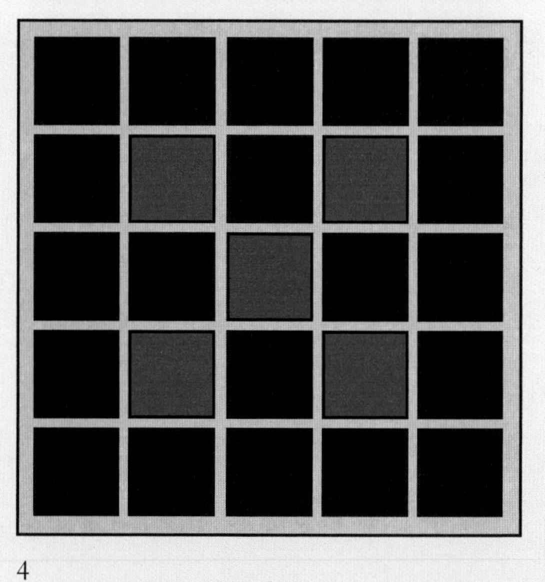

4

GALILEO'S PARADOX (PAGE 46)

Galileo's paradox exemplifies just one of the surprising properties of infinite sets. In his final work *Two New Sciences*, Galileo made two apparently contradictory statements. The set of squares and nonsquares is seemingly more numerous than that of just squares, but for every number there is exactly one square, and for every square there is exactly one number that is its square root, and therefore there cannot be more of one than the other. This was an early use of a proof by one-to-one correspondence.

CANTOR'S COMB (PAGE 47)

The formula for the total distance covered by the teeth at the nth stage of Cantor's comb is $(2/3)^n$. As n approaches infinity, Cantor's set approaches zero. A remarkable property of Cantor's comb is that at every stage of the comb, for any value between 0 and 1, you can always find two points on the comb whose horizontal distance apart is equal to that value.

EIGHT ONE OUT (PAGE 48)

All are prime numbers except the last one, which is the product of 17 and 19,607,843.

HOTEL INFINITY (PAGE 49)

In this case the hotel manager moves everyone to the room whose number is double their original number.

An infinite number of rooms will be vacated to accommodate the infinite number of arriving guests.

This paradox is called Hibbert's Hotel, named after the German mathematician David Hibbert (1862–1943). Essentially, it says that (2 × infinity) is still infinity.

PRIME NUMBERS (PAGE 50)

A prime number is a number that is only divisible by itself and 1. Among the three numbers the only prime is 2.

The other two numbers can be factored as shown:

$117 = 9 \times 13$

$539 = 7 \times 77$

▼ SIEVE OF ERATOSTHENES (PAGE 51)

Removing the multiples of the first few primes (2, 3, 5, and 7) is sufficient to find all the primes smaller than 100. The multiples of 11 and higher primes are not required since, for example, the composite nature of 77 = 7 x 11 has already been accounted for by the multiples of 7. If you think about it, you will realize that in order to find the primes from 1 to x, you need only remove the multiples of primes that are equal to or less than the square root of x.

In this example, we only needed multiples of primes equal to or less than the square root of 100—namely 2, 3, 5, and 7.

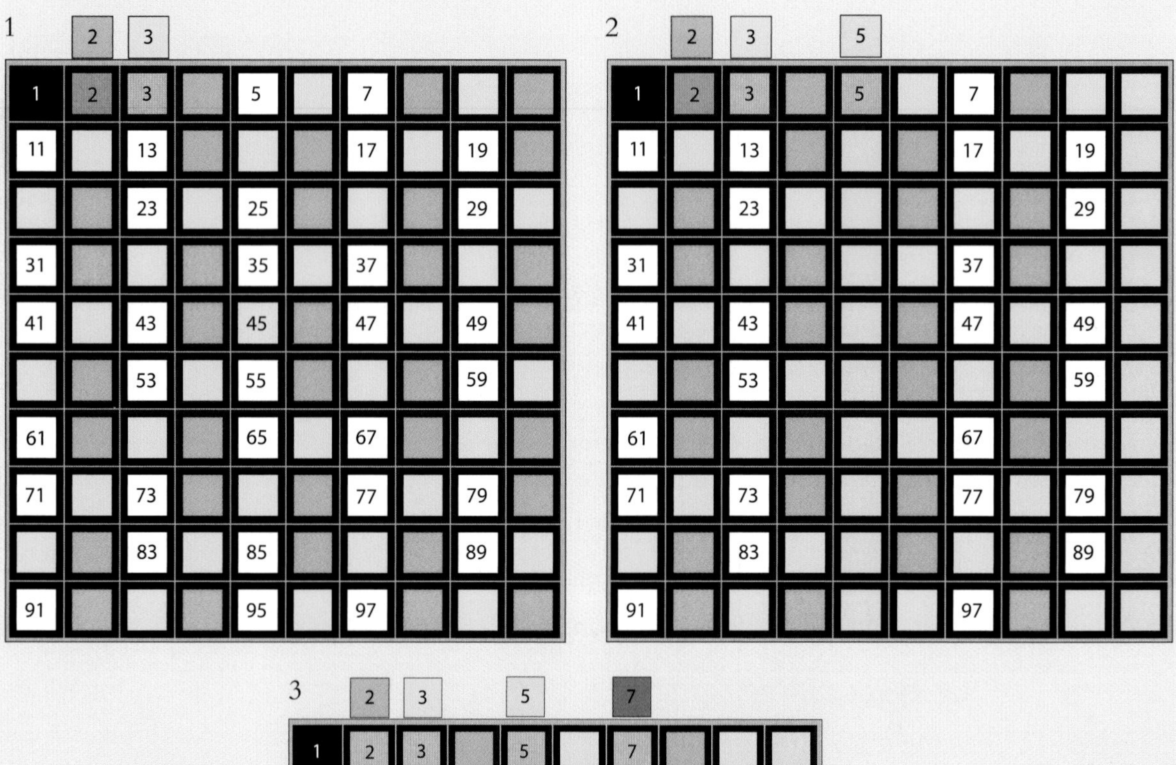

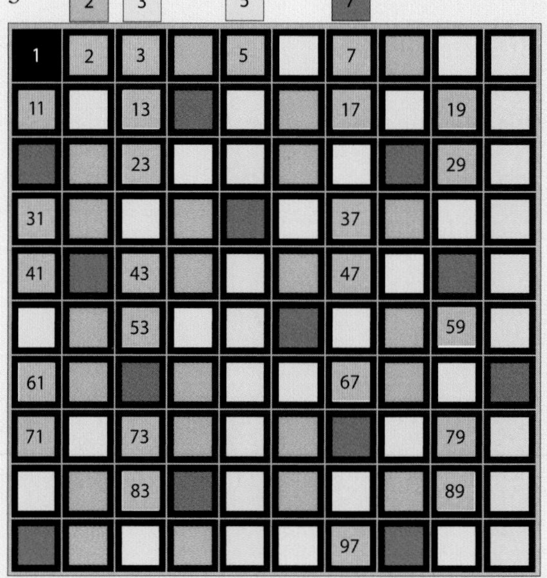

ALL NINES PARADOX (PAGE 52)

Of the first 1000 numbers, 271 have a 9 in them, which is about 27% of the total. Surprisingly enough, 99% of the numbers included in the course of counting to our large number (10^{64}), have a 9 in them. This may lead us to think that almost all numbers have a 9 in them!

But there is nothing special about 9. For every number with a 9 in it, there is a corresponding number with an 8 in the same position (or 7, 6, 5, 4, 3, 2, or 1), so almost all numbers contain every digit!

▼ NUMBER PATTERN (PAGE 54)

Each number not along the top or left borders is the sum of the number above it and the number to its left, minus the number diagonally above it to the left.

1	2	5	6	9
3	4	7	8	11
10	11	14	15	18
12	13	16	17	20
19	20	23	24	27

PUNCHED CARDS PUZZLE (PAGE 55)

Stack the punched cards as shown to reveal different colors in all blank windows.

▼ PRIME SPIRAL (PAGE 57)

In this spiral, one of the main diagonals consists entirely of prime numbers, as shown.

213	212	211	210	209	208	207	206	205	204	203	202	201	200	199
214	161	160	159	158	157	156	155	154	153	152	151	150	149	198
215	162	117	116	115	114	113	112	111	110	109	108	107	148	197
216	163	118	81	80	79	78	77	76	75	74	73	106	147	196
217	164	119	82	53	52	51	50	49	48	47	72	105	146	195
218	165	120	83	54	33	32	31	30	29	46	71	104	145	194
219	166	121	84	55	34	21	20	19	28	45	70	103	144	193
220	167	122	85	56	35	22	17	18	27	44	69	102	143	192
221	168	123	86	57	36	23	24	25	26	43	68	101	142	191
222	169	124	87	58	37	38	39	40	41	42	67	100	141	190
223	170	125	88	59	60	61	62	63	64	65	66	99	140	189
224	171	126	89	90	91	92	93	94	95	96	97	98	139	188
225	172	127	128	129	130	131	132	133	134	135	136	137	138	187
226	173	174	175	176	177	178	179	180	181	182	183	184	185	186
227	228	229	230	231	232	233	234	235	236	237	238	239	240	241

BINARY ABACUS (PAGE 59)

1) 1023
2) 170
3) 181
4) 871
5) 585

▶ BINARY WHEELS (PAGES 60–61)

The 3-bit solution is unique.

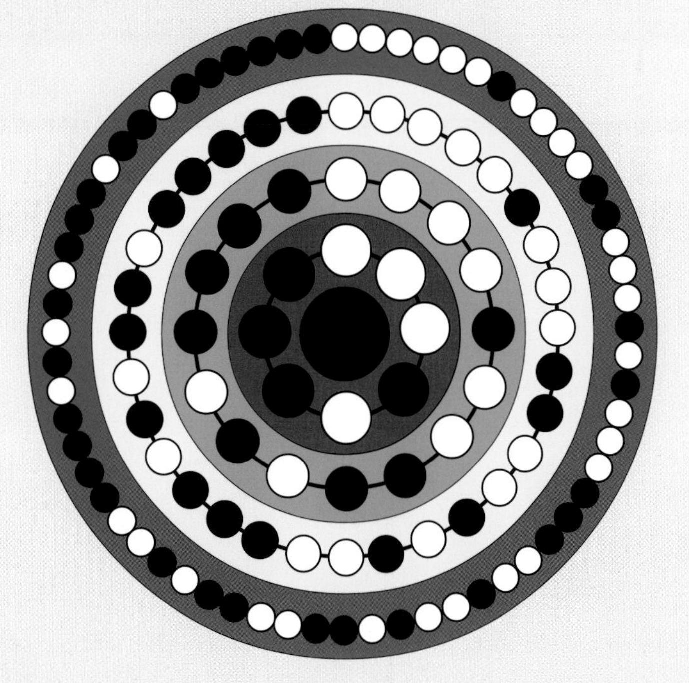

▼ Q-BITS GAME (PAGE 62)

The shortest possible two-person game ends in eight moves, and there are many solutions, one of which is shown. As you can see, none of the remaining eight tiles can be fitted in the board, because touching corners are of opposing colors.

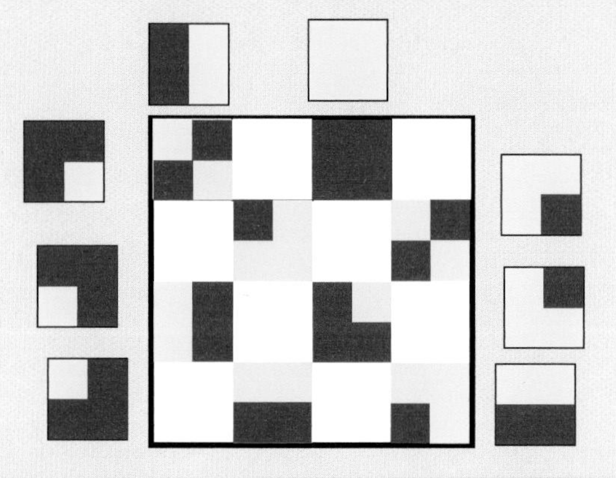

▼ Q-BITS GRIDS (PAGE 64)

Here are the 50 possible solutions.

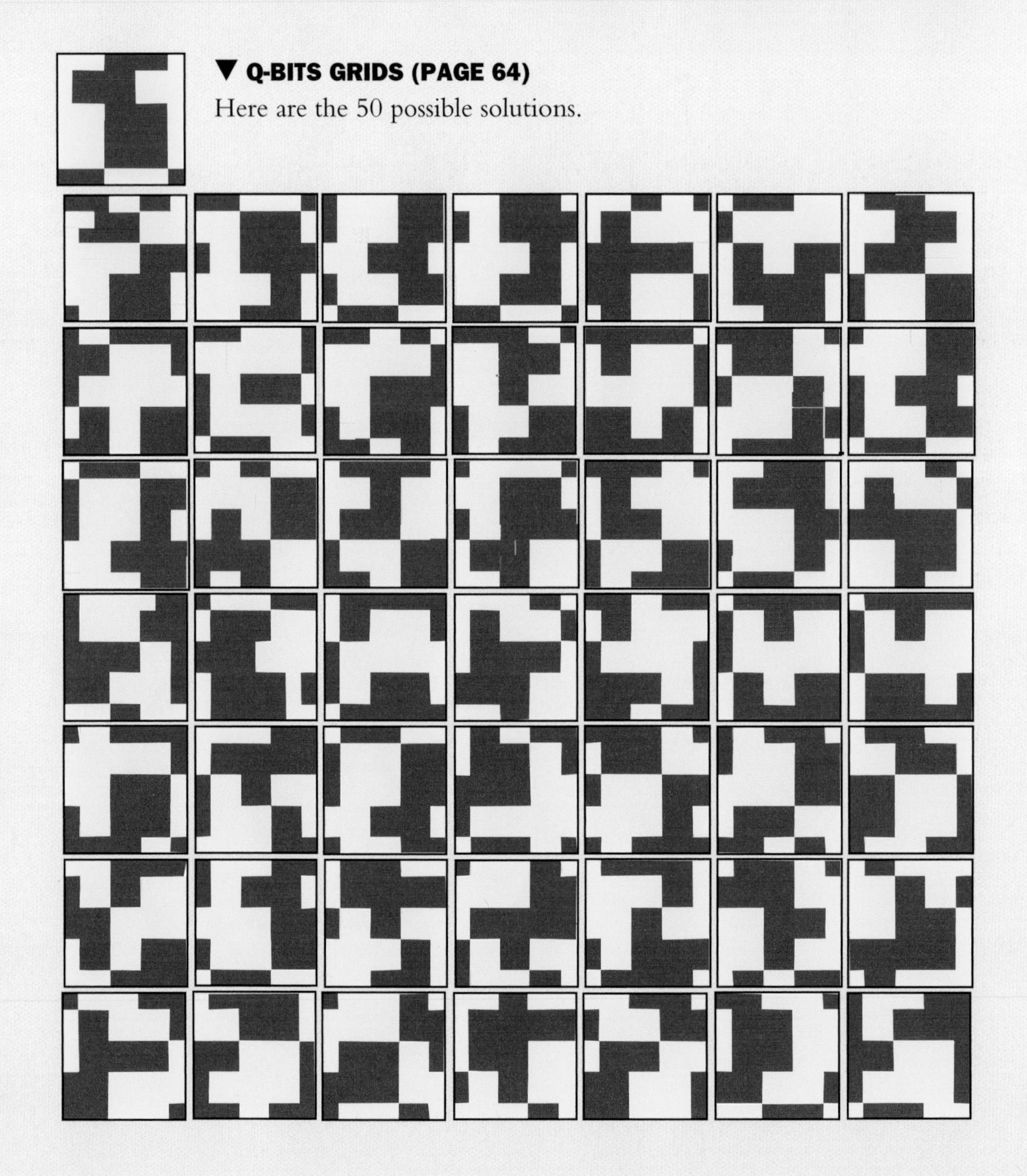

▼ FROGS AND PRINCES (PAGE 65)

The secret is to look at the eight tiles in the shaded areas shown below. If there is an even number of frogs and princes in the shaded squares the game has a solution—otherwise it doesn't! The reason this is so is because it is impossible to make any move that affects an odd number of those squares; every move will flip either zero or two of them. Since you must end the puzzle with all eight of the tiles on those squares showing the same face, the puzzle cannot be solved if there is an odd number of princes and frogs on those squares. Accordingly, the first configuration is unsolvable, while the second is solvable.

▼ GLASSES UP (PAGE 66)

Three moves are necessary, as shown.

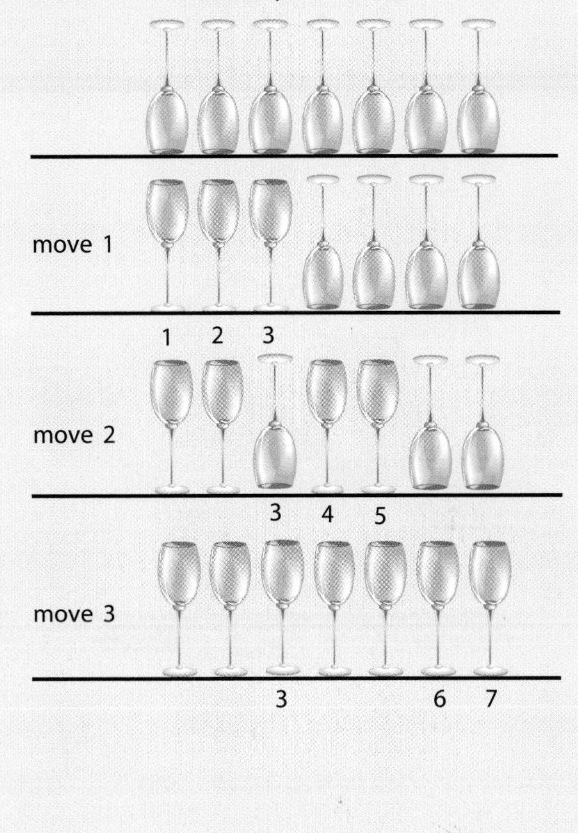

TEN GLASSES PROBLEM (PAGE 67)

The parity of the glasses is odd. It cannot be changed to an even parity by an even number of moves.

The term "parity" was first used in math to distinguish between even and odd numbers. If two numbers are both even or both odd, they have the same parity; otherwise they have opposite parity. The parity of a pattern is conserved by an even number of moves.

▼ BINARY TRANSFORMATIONS (PAGE 68)

The numbers along the sides of the grids indicate the solutions. First, exchange the vertical columns so that, instead of the scrambled order shown, they read 1–6 in numerical order. Then rearrange the horizontal rows in the same way.

Note how each transformation was achieved using exactly the same series of exchanges.

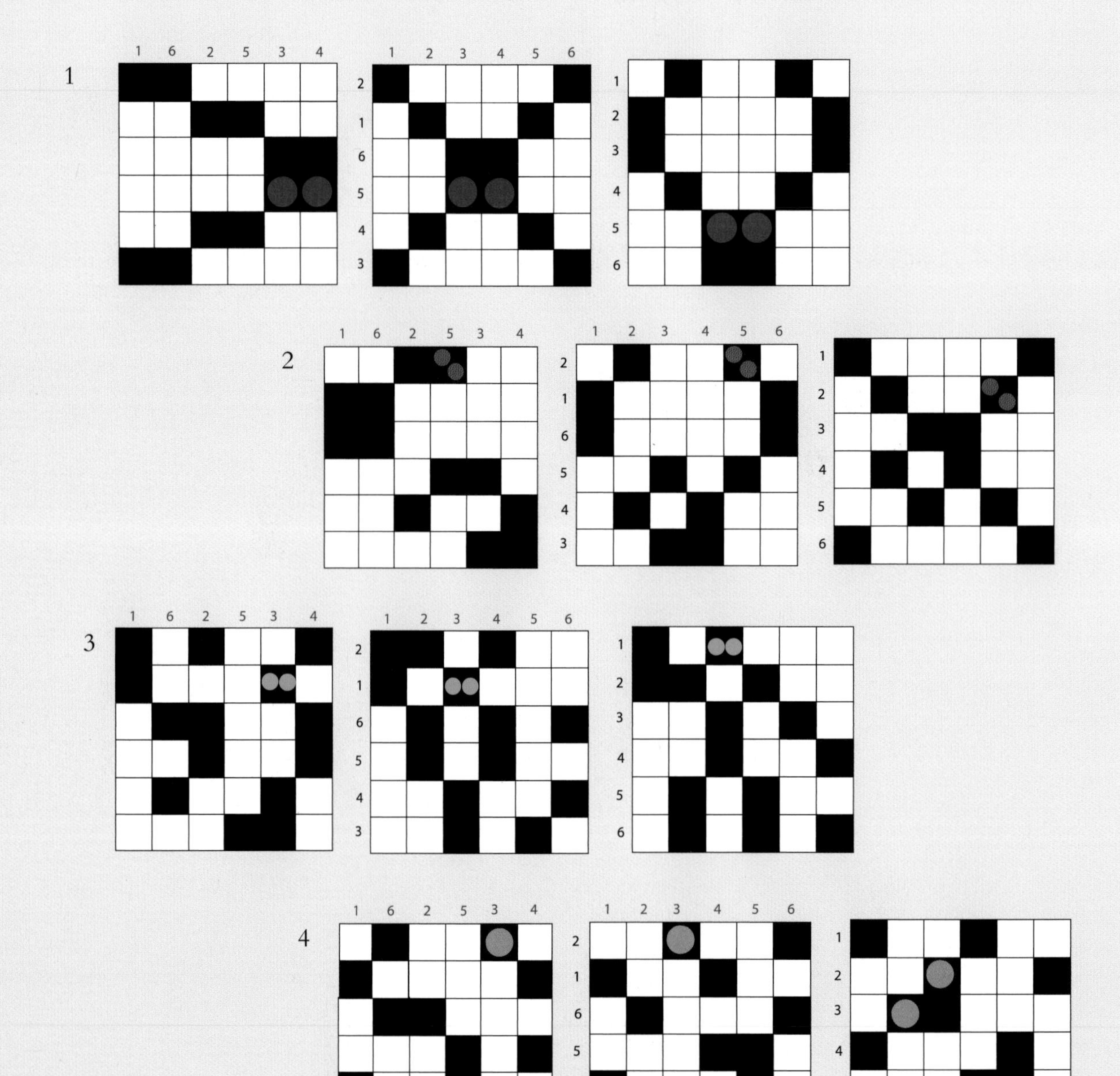

▼ BINARY TRANSFORMATIONS (PAGE 69)

5

6

7

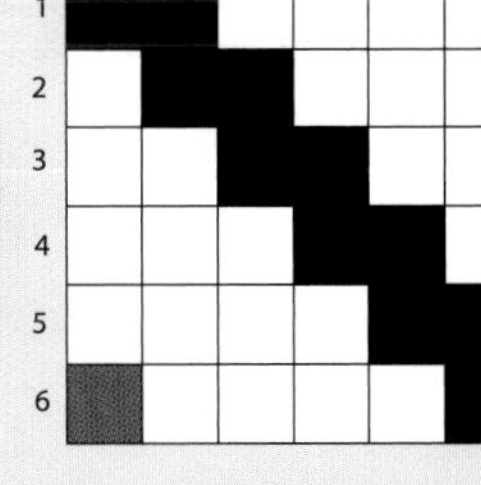

8

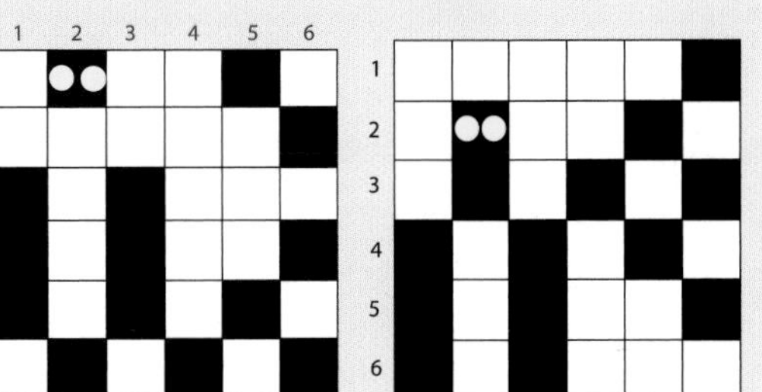

9

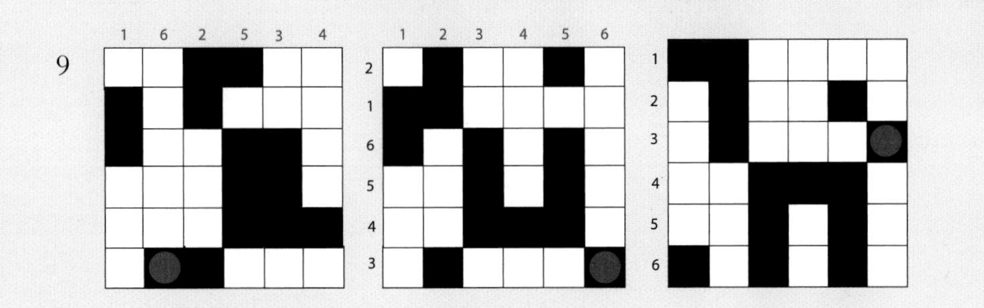

TWO FATHERS (PAGE 70)

$1 \times 1 \times 3 \times 13 = 39$

FATHER AND SON (PAGE 71)

The father could be 96 and the son 69, or the ages could be 85 and 58, 74 and 47, 63 and 36, or 52 and 25. But judging from the picture, it is more likely that the father is 41 and the son 14.

LOST MARBLES (PAGE 72)

40 marbles.

Let the number of marbles at the start be x. There are 2x at the start. $2x + 35 - 15 = 100$, so $2x + 20 = 100$ or $2x = 80$, so $x = 40$.

ARISTOTLE'S WHEEL PARADOX (PAGE 73)

As for the first question, points near the top have faster ground speeds than points at the bottom. On flange train wheels, whose rims extend below the track, there are even points which move backwards.

The paradox went unexplained for many years until George Cantor solved it in 1869.

The fallacy lies in the assumption that a one-to-one correspondence between the two wheels means that two curves must have the same length. In fact they don't. The big wheel does roll from point 1 to point 2, but the smaller wheel does not roll from 3 to 4. It is dragged along the line.

All points on a one-inch segment can be put in one-to-one correspondence with all points on a line a mile long as well as on a line of infinite length. The number of points on any segment of a curve is what Cantor called aleph-one, the second of his transfinite numbers.

Mathematicians before Cantor were not familiar with the strange properties of transfinite numbers, so all their attempts to solve Aristotle's wheel paradox were doomed to failure.

LITTLE WOODEN MAN (PAGE 74)

He does not jump at all, since he doesn't hear the clock strike. He's made of wood! I warned you it was tricky.

▶ INTEGRAL RECTANGLE (PAGE 75)

Any large rectangle constructed this way must have at least its width or its height be integral, and possibly both. This was proven by Stan Wagon in his article "Fourteen Proofs of a Result about Tiling a Rectangle."

Peter Winkler described his ingenious proof in his book *Mathematical Puzzles: A Connoisseur's Collection*.

Color each rectangle of integral width green with a thick orange line on top and bottom. Color the rest orange with a thick vertical green line left and right.

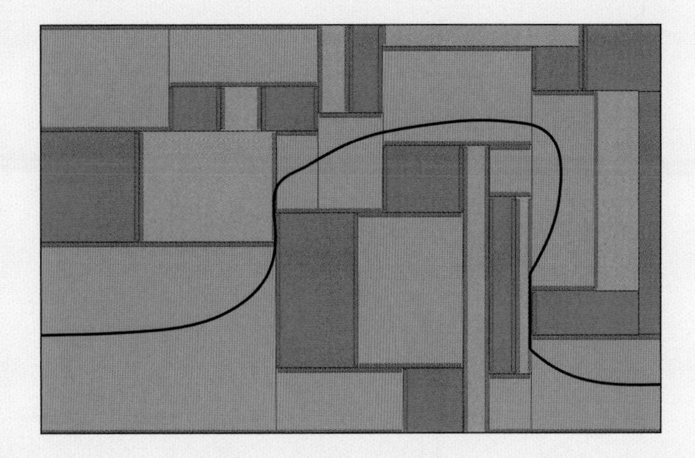

With any rectangle constructed in this fashion, there must be at least one path connecting opposite sides of the rectangle—either a green path from the left side of the big rectangle to the right side, or an orange path from top to bottom. (Colored areas that meet at a corner are considered to be adjacent, so it's possible to have two paths that cross each other.) We can see from the diagram at right that our rectangle has integral width only.

Try this with your own rectangle diagrams!

▼ ANIMAL PROMENADE (PAGE 76)

Starting from the lower left corner, the four animals spiral in a counterclockwise direction in the same sequence, as shown.

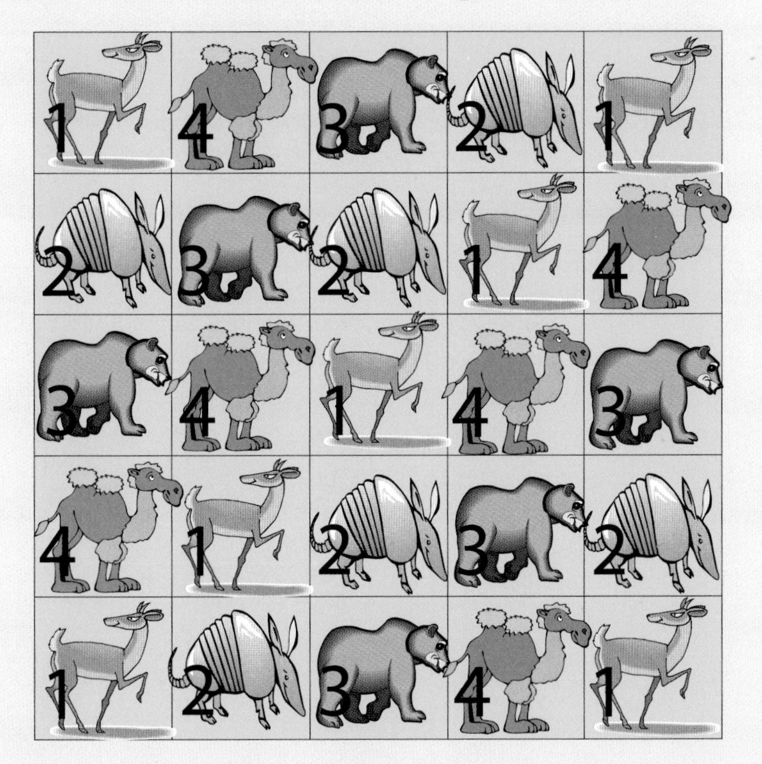

▼ WHAT'S IN THE SQUARE? (PAGE 77)

A red square is missing somewhere in the third horizontal row.

SEVEN BIRDS (PAGE 78)

Day 1: 1 2 3
Day 2: 1 4 5
Day 3: 1 6 7
Day 4: 2 4 6
Day 5: 2 5 7
Day 6: 3 4 7
Day 7: 3 5 6

▶ WALKING DOGS (PAGE 79)

First work out how many pairings are possible among the 9 girls.

There are 36 pairs, shown in the red box at right.

Three pairs are involved in each triplet, so each pairing will be used once in the 12 triplets (three triplets over four days). Following a systematic procedure you can find a sequence of triplets that fulfills the conditions (see chart below).

Day 1	123	456	789
Day 2	147	258	369
Day 3	159	263	348
Day 4	168	249	357

1 - 2
1 - 3
1 - 4
1 - 5
1 - 6
1 - 7
1 - 8
1 - 9
2 - 3
2 - 4
2 - 5
2 - 6
2 - 7
2 - 8
2 - 9
3 - 4
3 - 5
3 - 6
3 - 7
3 - 8
3 - 9
4 - 5
4 - 6
4 - 7
4 - 9
5 - 6
5 - 7
5 - 8
5 - 9
6 - 7
6 - 8
6 - 9
7 - 8
7 - 9
8 - 9

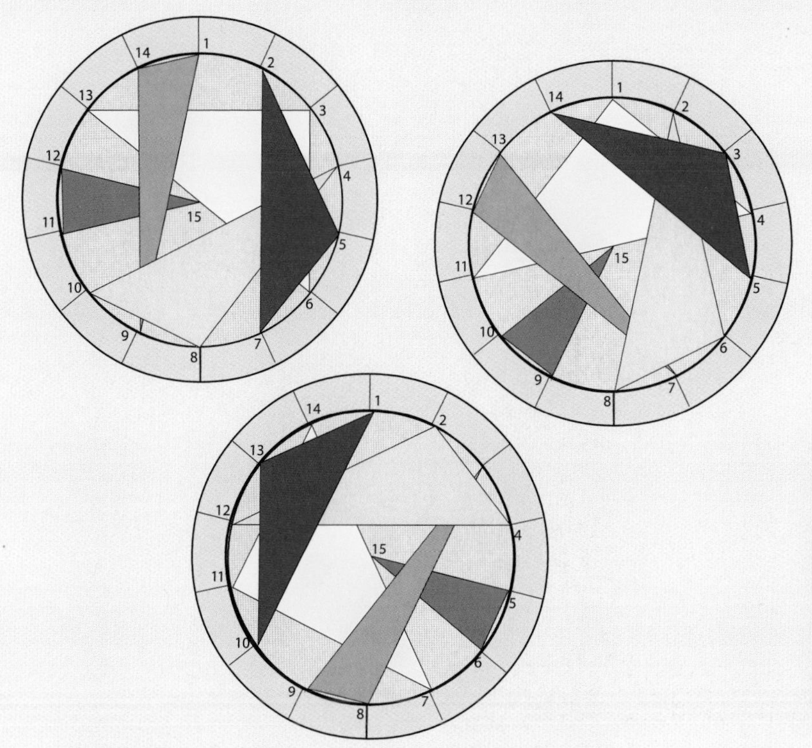

◀ SCHEDULING SCHOOLCHILDREN (PAGE 80)

Ingenious geometric methods have been devised to solve problems like this. One is demonstrated. Around the outer disk 14 points are distributed equally. The inner rotating wheel, with a pattern of colored triangles, is rotated about the center point, numbered 15. It is rotated two units at a time to seven different positions to provide the seven sets of triplets, as shown below.

TRIPLE S

Day 1	1 2 15	3 7 10	4 5 13	6 9 11	8 12 14
Day 2	1 5 8	2 3 11	4 7 9	6 10 12	13 14 15
Day 3	1 9 14	2 5 7	3 6 13	4 8 10	11 12 15
Day 4	1 4 11	2 6 8	3 5 14	7 12 13	9 10 15
Day 5	1 3 12	2 9 13	4 6 14	5 10 11	7 8 15
Day 6	1 10 13	2 4 12	3 8 9	5 6 15	7 11 14
Day 7	1 6 7	2 10 14	3 4 15	5 9 12	8 11 13

		CCCMMM	
	◀ C C	CMMM	1
C	● C ▶	CMMM	2
C	◀ C C	MMM	3
CC	● C ▶	MMM	4
CC	◀ M M	CM	5
C M	C M ▶	C M	6
CM	◀ M M	CC	7
MMM	● C ▶	CC	8
MMM	◀ C C	C	9
MMMC	● C ▶	C	10
MMMC	◀ C C		11

Ⓒ cats Ⓜ mice

◀ CATS AND MICE CROSSING (PAGE 81)

There are four different minimal solutions, all requiring 11 moves. One of them is shown at left.

▶ PINWHEEL PATTERN (PAGE 82)

The pattern is based on the number of possible permutations of four elements, in our case four colors. There are 24 different permutations of four colors; in our puzzle those colors are arranged in a cyclic configuration, reducing the number of different pinwheels to only six, not considering rotations to be different. Every horizontal row and vertical column contains the six different cyclic permutations (given below, reading clockwise from yellow):

1) yellow-red-green-blue
2) yellow-red-blue-green
3) yellow-green-red-blue
4) yellow-green-blue-red
5) yellow-blue-red-green
6) yellow-blue-green-red

The blank pinwheels may be colored in any rotation of the appropriate permutations.

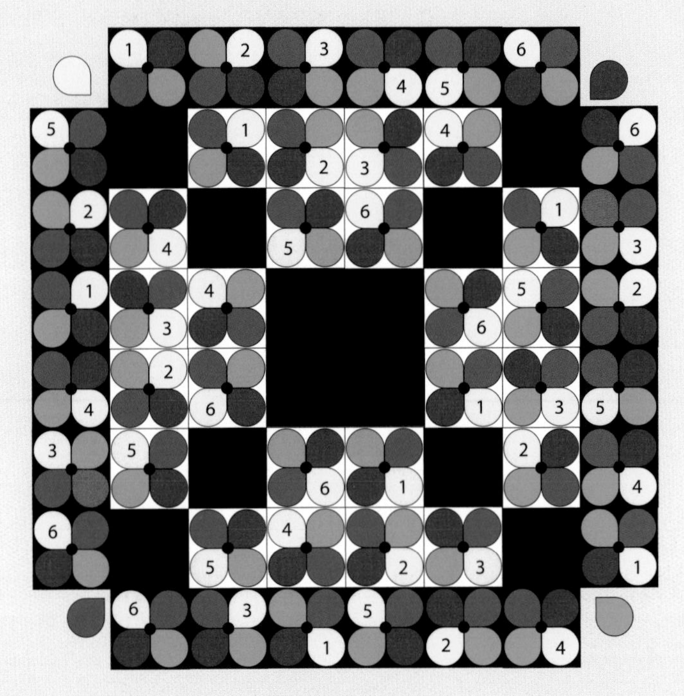

WHERE IS TOM? (PAGE 83)

Tom is on the right. Harry is in the middle. Dick is on the left, and is lying.

TRUE STATEMENT (PAGE 84)

Statement 11 must be true. Each of the 12 statements contradicts the other 11, so only one of them can be true; thus 11 of them are false.

TRUTH CITY ROAD (PAGE 85)

Ask the man: "Please point to the road leading to the city you are from." If the man is from Truth City, he will point to its road. If he is from the city of Liars, he will also point to the Truth City.

The interesting thing about his answer is that although you have obtained the information you needed, you don't know whether the man told the truth or lied.

(The liar and truthteller puzzles in this book were inspired by Raymond Smullyan.)

TRUTH, LIES, AND IN BETWEEN (PAGE 86)

Ask the man this question twice: "Are you one of those who alternately lies and tells the truth?"

Two "no" answers indicates he is a truthteller.

Two "yes" answers indicates he is a liar.

If he gives two different answers, then he is someone who alternately lies and tells the truth.

TRUTH AND MARRIAGE (PAGE 87)

The young man should ask one of the daughters, "Are you married?"

Regardless of which daughter is asked the question, a "yes" answer means Amelia is married, while a "no" means Leila is married.

For example, if Amelia was the one the question was directed to, an answer of "yes" would be the truth—meaning she is married. If she answers "no," then she is truthfully saying she is not married, so Leila must be.

On the other hand, if the question was asked of Leila, an answer of "yes" would be a lie, indicating she is not married, and Amelia must be. If she answers "no," she is lying, and is actually married.

So even though the young man doesn't know which daughter is which, by asking this question he will be able to tell the king the name of the unmarried daughter.

SWIMMING POOL (PAGE 88)

Yes. The swimming pool contains about 50 cubic meters of water, which is 50,000 liters of water.

For health reasons, we are advised to drink at least 2 liters of water per day, which is about 730 liters per year.

It would take you about 68 years to drink the water in the swimming pool.

FIGARO (PAGE 89)

Figaro has a beard. Of all the men who do not have a beard, every one of them either shaves himself or has Figaro shave him. Furthermore, no man adopts any mixed policy of shaving himself and having Figaro shave him as well. But in Figaro's case this means that he never shaves himself at all. For if he did, he would be shaving himself, and be shaved by Figaro. No man does that. Therefore, Figaro has a beard.

HATS AND PRISONERS (PAGE 90)

All the prisoners can be set free if they line up correctly.

The first prisoner starts the line. The others insert themselves into the line behind the last red hat they can see (or in front of the first black hat they can see). This will produce a line with all the red hats in front and all the black hats in the rear. Since the new prisoner is always in the middle (between red and black), he will know his color the moment the next prisoner joins the line.

If the new prisoner joins the line in front of him, then he has a black hat. This takes care of 99 prisoners.

So when the last prisoner joins the line, the one standing in the front simply leaves his position and re-inserts himself between red and black . All the 100 prisoners are saved!

▼ MY CLASS (PAGE 91)

As can be seen from the table below, there can be as many as 10 such boys or as few as one. So at least one boy must have all four features.

1	2	3	4	5	6	7	8	9	10	11	12	13	14	15	16	17	18	19	20
					blue eyes														
				black hair															
				overweight															
				tall															

1	2	3	4	5	6	7	8	9	10	11	12	13	14	15	16	17	18	19	20
					blue eyes														
												black hair							
				overweight															
														tall					

COCKTAIL AT LARGE (PAGE 92)

The circumference.

▼ HOTEL DOORS (PAGE 93)

Doors 1, 4, and 9 will be closed at the end. Door N is altered at step K if and only if N is divisible by K. A door ends up closed or open according to the parity (evenness or oddness) of the number of times the door's state changes. Square numbers have a different parity than other numbers. Non-squares have an even number of factors (for instance, the factors of 10 are 1, 2, 5, and 10), but perfect squares have an odd number of factors (the factors of 9 are 1, 3, and 9). Do you see why this is?

step K	door N									
	1	2	3	4	5	6	7	8	9	10
1										
2		●		●		●		●		●
3			●			●			●	
4				●				●		
5					●					●
6						●				
7							●			
8								●		
9									●	
10										●

HATS AND COLORS 1 (PAGE 94)

Clown B.

If clown A saw two red or two green hats he would know the color of his hat (as well as the color of clown D's hat). But what he sees is one of each color, which doesn't offer him any clues.

Hearing that clown A is silent, clown B can deduce that the color of his hat must be the opposite color of the hat in front of him.

HATS AND COLORS 2 (PAGE 95)

Clown A sees two red hats and one blue hat. His hat can be either red or blue.

Clown B knows that A sees only one blue hat. Therefore, he can deduce that the color of his hat is red.

Clown C can't know the color of his hat.

But the question was to determine who could deduce the color of clown A's hat.

Only clown D can do this. He knows that clown A sees neither two blue hats (or he would say his hat is red) nor three three red hats (or he would know his hat is blue). Therefore, clown D knows clown A can see two red hats and one blue hat, leaving one red hat and one blue hat to be worn by clowns A and E in some order. Since clown D can see clown E's hat, clown D knows that clown A is wearing the color of hat that clown E is not wearing.

▼ ALHAMBRA PATTERN

The pattern is composed of 25 interlocking closed loops of three shapes in different orientations as shown.

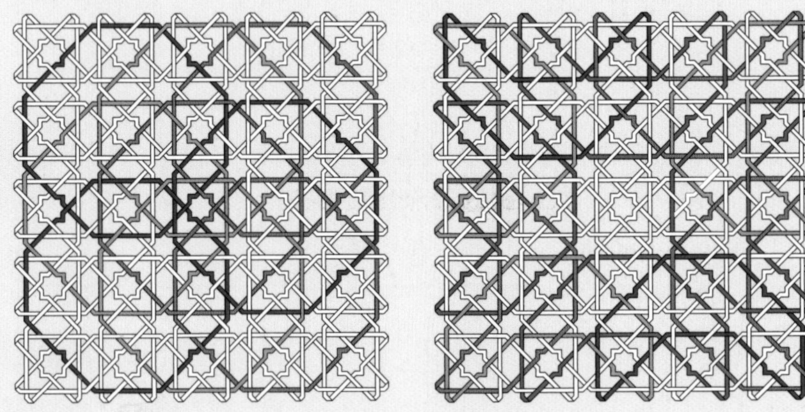

Shape 1: nine identical shapes

Shape 2: twelve identical shapes in four different orientations

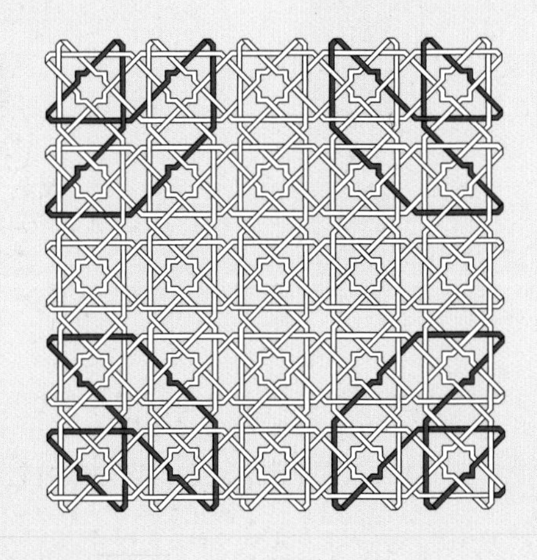

Shape 3: four identical shapes in four different orientations

ODD ONE OUT (PAGE 97)

Second from left.